AS Fast-Track

Maths

Lynn Byrd
Paul Dexter
Daniel Dryer
Dee Spackman

Series consultants **Geoff Black and Stuart Wall**

Page designer **Michelle Cannatella**

Cover designer **Kube Ltd**

Pearson Education Limited
Edinburgh Gate
Harlow
Essex CM20 2JE, England
and Associated Companies throughout the world

ISBN 0 582 43237-5

British Library Cataloguing-in-Publication Data

A catalogue record for this book is available from the British Library.

Set by 3 in Optima and Tekton
Printed by Ashford Colour Press, Gosport, Hants

Contents

Read this first!

TWO WAYS TO USE THIS BOOK...

This book is designed to be used:

Either

- On its own – work through all the exercises for a quick run-through of your subject. This will take you about 24 hours altogether.

or

- As part of the *Revision Express* system:

1 Read through the topic in the *Revision Express A-level Study Guide* (or similar book).

2 Work through the exercises in this book.

3 Go to www.revision-express.com for extra exam questions and model answers.

4 For even more depth and detail, refer back to your textbook or class notes, and visit the web links from www.revision-express.com.

HOW THE BOOK WORKS

The book is divided into two-page revision sessions. To make your revision really effective, study one session at a time.

Have a short break between sessions – that way you'll learn more!

Each session has two parts:

1st page: the first page on each topic contains interactive exercises to nail down the basics. Follow the instructions in the margin and write your answers in the spaces provided.

2nd page: the second page contains exam questions. Sometimes you'll answer the exam question directly, but more often you'll use it as a starting point for in-depth revision exercises. In each case, follow the extra instructions in the margin.

REMEMBER: the answers in the back are for the revision exercises – they are not necessarily model answers to the exam questions themselves. For model answers to a selection of exam questions go to www.revision-express.com.

All the pages are hole-punched, so you can remove them and put them in your folder.

TRACK YOUR PROGRESS

The circles beside each topic heading let you track your progress.

If a topic is hard, fill in one circle. If it's easy, fill in all three. If you've only filled in one or two circles go back to the topic later.

TOPIC HEADING

EXAM BOARDS

You might not need to work through every session in this book. Check that your exam board is listed above the topic heading before you start a session.

AS AQA EDEXCEL OCR WJEC

This book covers the most popular topics. For full information about your syllabus, contact the relevant exam board or go to their website.

AQA
(Assessment and Qualifications Alliance)
Publications department, Stag Hill House, Guildford, Surrey GU2 5XJ – www.aqa.org.uk

EDEXCEL
Stuart House, 32 Russell Square, London WC1B 5DN – www.edexcel.org.uk

OCR
(Oxford, Cambridge and Royal Society of Arts)
1 Hills Road, Cambridge CB2 1GG –
www.ocr.org.uk

DON'T FORGET

Exam questions have been specially written for this book. Ask your teacher or the exam board for the official sample papers to add to the questions given here.

COMMENTS PLEASE!

However you use this book, we'd welcome your comments. Just go to www.revision-express.com and tell us what you think!

GOOD LUCK!

Algebra – the language of mathematics

You need confidence to cope with all aspects of algebra at AS-level, so give yourself lots of practice.

MANIPULATION OF FORMULAE ○○○

Rearrange these formulae to make the letter in the bracket the subject – take it one step at a time and show all working.

DON'T FORGET
You can use the symbol $\therefore$ (therefore).

$y = 2x + 3 \ (x)$

$v^2 = u^2 + 2as \ (s)$

$V = \pi r^2 h \ (r)$

$y = \dfrac{x + 1}{x - 1} \ (x)$

EXPANDING AND FACTORIZING ○○○

Here are some examples of taking brackets out and putting brackets in. Highlight the correct answer and write its letter in the box.

$5x(x - 7)$	= (A) $5x^2 - 12x$	(B) $5x^2 + 35x$	(C) $5x^2 - 35x$	☐
$6x^2 + 8xy$	= (A) $2x(4x + 4y)$	(B) $2x(3x + 4y)$	(C) $2x^2(3 + 4y)$	☐
$(3x - 5)(2x + 7)$	= (A) $5x^2 + 11x + 2$	(B) $6x^2 + 11x - 35$	(C) $6x^2 - 11x - 35$	☐
$2x^2 - x - 6$	= (A) $(2x + 3)(x - 2)$	(B) $(2x - 3)(x + 2)$	(C) $(2x - 3)(x - 2)$	☐

SOLVING QUADRATIC EQUATIONS ○○○

Equations of the type $ax^2 + bx + c = 0$ are called quadratic equations, and there are three ways to solve them. Get used to using all three – don't just use your favourite.

Solve these equations by factorizing.

DON'T FORGET
After factorizing, check your factors are correct by multiplying out the brackets. You should obtain the expression you started with.

$x^2 + 3x - 10 = 0$

$2y^2 + 3y - 20 = 0$

$3t(t + 2) + 3(t - 1) = t$

Write the formula here.

EXAMINER'S SECRETS
Always quote the formula first.

Solve these two equations using the formula. Leave the solutions to the first equation in surd form and give the solutions to the second equation correct to 2 d.p.

Now solve these equations.

EXAMINER'S SECRETS
In the exam you probably won't be told which method to use, so if the question asks for your answer to 2 d.p., that's a big hint to use the formula.

$x^2 - 3x + 1 = 0$

$2x^2 + 3x - 1 = 0$

Turn the page for some exam questions on this topic ➤

EXAM QUESTION 1

Find the roots of the equation $2(x + 1)(x - 2) + x(x - 3) = 5$, leaving your answer in surd form.

Write the phrase in the question which suggests you need to use the formula.

Now have a go at solving the equation.

DON'T FORGET
Quote the formula first.

EXAM QUESTION 2

(a) Expand $(1 + x)(1 + x + x^2)$ in ascending powers of x.

(b) Find as a decimal the exact value of $(1 + x)(1 + x + x^2)$ for $x = 10^{-4}$.

Part (a) is straightforward but take your time and show every step. For part (b), think back to your GCSE standard form.

DON'T FORGET
$a^m a^n = a^{m+n}$
$(a^m)^n = a^{mn}$
$a^m/a^n = a^{m-n}$

EXAMINER'S SECRETS
The phrase 'find the exact value' is a big hint that you may have do some algebra before you use your calculator.

EXAMINER'S SECRETS
Read the question carefully. It asks you to find the 'values', and this suggests there is more than one answer.

EXAM QUESTION 3

Find the values of x which satisfy the equation
$$\frac{3}{x+3} - \frac{x}{x+2} = 1$$

To subtract fractions, first find a common denominator. To practise try $\frac{2}{3} - \frac{1}{2}$.

Algebra – quadratics

We've solved quadratic equations by formula and factorization. The third option is completing the square, a useful way to find the minimum or maximum value of a quadratic function.

COMPLETING THE SQUARE ○○○

Complete the square for these quadratic functions.

$x^2 + 4x - 7 \quad = (x + \quad)^2 - 7 - \quad = (x - \quad)^2 -$

$x^2 - 10x + 3 \quad = (x - \quad)^2 + 3 - \quad = (x - \quad)^2 -$

$-2x^2 - 12x - 8 = -2(\qquad\qquad) = -2((x + \quad)^2 + 4 - \quad)$

$\qquad\qquad = -2((x + \quad)^2 - \quad) = -2(x + \quad)^2 +$

$4x^2 + 14x - 2 \quad = (2x + \quad)^2 - 2 - (\quad)^2 = (2x + \quad)^2 -$

Putting these quadratic functions equal to zero, we have a quadratic equation which we can now solve using our completed square.

Finish solving these equations using your answers from above. Leave your answers in surd form.

$x^2 + 4x - 7 = 0 \quad \therefore (x + 2)^2 - 11 = 0$

$x^2 - 10x + 3 = 0 \quad \therefore (x - 5)^2 - 22 = 0$

$-2x^2 - 12x - 8 = 0 \text{ (divide by } -2) \; x^2 + 6x + 4 = 0$

$4x^2 + 14x - 2 = 0 \quad \therefore (2x + 3\frac{1}{2})^2 - 14\frac{1}{4} = 0$

Once we've completed the square on a quadratic function we can also state the maximum or minimum value of the function by looking at the constant at the end. Just remember the smallest value for the ( )^2 term is zero.

Use your answers from above to find max or min values. The first one has been done for you.

$x^2 + 4x - 7 \quad = (x + 2)^2 - 11: \text{min} -11 \text{ when } x = -2$

$x^2 - 10x + 3 \quad =$

$-2x^2 - 12x - 8 =$

$4x^2 + 14x - 2 \quad =$

THE DISCRIMINANT ○○○

The discriminant $b^2 - 4ac$ tells us about the roots of a quadratic. Draw a line from each inequality to its correct root statement.

1 $b^2 - 4ac < 0$ one repeated root

2 $b^2 - 4ac > 0$ no real roots

3 $b^2 - 4ac = 0$ two distinct roots

Find $b^2 - 4ac$ for these quadratics and describe their roots.

$x^2 + 3x + 10 = 0 \quad b^2 - 4ac =$

$6 - 2x - x^2 = 0 \quad b^2 - 4ac =$

$x^2 - 20x + 25 = 0 \quad b^2 - 4ac =$

Turn the page for some exam questions on this topic ➤

EXAM QUESTION 1 ● ● ●

Part (a) is the examiner's way of asking you to complete the square. In part (b) the word 'hence' means you should use what you've already worked out.

The function $f(x) = 2x^2 + 12x - 6$ can be written in the form $p(x + q)^2 - r$, where p, q and r are positive real numbers. (a) Find the values of p, q and r. (b) Hence find coordinates of the minimum point on the graph of $f(x)$. (c) Find the coordinates of the points where the curve crosses the x-axis; leave your answers in surd form. (d) Sketch the graph.

DON'T FORGET
When completing the square on a function, you can only factor out a number (rather than divide through by the number) as the function is not yet an equation.

DON'T FORGET
If you are asked for coordinates, give the (x, y) coordinates.

EXAM QUESTION 2 ● ● ●

(a) By substituting $y = x^{1/2}$ into the equation $4x = 13x^{1/2} - 3$, obtain a new equation in y. (b) Solve the equation to find the values of y. (c) Hence find the values of x.

EXAMINER'S SECRETS
When working with fractions, it's often a lot easier to use top-heavy fractions rather than mixed numbers.

DON'T FORGET
Methods mean marks, mmm!

EXAM QUESTION 3 ● ● ●

The curve C has the equation $y = 3x - 4 - x^2$. Show algebraically that the graph of C does not cross the x-axis.

Algebra – simultaneous equations

To find the coordinates of the point where two lines intersect, or the point(s) where a line and a curve intersect, we need to solve the equations of the lines simultaneously.

DON'T FORGET
Always start by numbering your equations, then decide to get either the xs or ys the same.

SOLVING TWO LINEAR EQUATIONS BY ELIMINATION ○○○

Find the coordinates of the point of intersection of the lines $4x + 3y = 11$ and $3x - 2y = 21$.

Decide the multipliers to get the y's the same. Write your new equations here, number them (3) and (4)

$$4x + 3y = 11 \quad (1)$$
$$3x - 2y = 21 \quad (2)$$

Now that the ys are the same you can add equations (3) and (4) to eliminate the ys, and find the value of x.

Show the next steps here.

The last part (often forgotten) is to substitute the x-value into one of the first two equations to find the y-value.

Substitute $x = 5$ into (1) to find the y-value.

EXAMINER'S SECRETS
Always finish by stating your final answer as a coordinate; you may need to do this to get your final mark.

SOLVING ONE LINEAR AND ONE QUADRATIC EQUATION BY ○○○ SUBSTITUTION

Find the coordinates of the points where the line $y + 5 = x$ intersects the curve $x^2 + xy = 3$.

$$y + 5 = x \quad (1)$$
$$x^2 + xy = 3 \quad (2)$$

Always look at the quadratic first (equation 2) and decide which variable it would be easiest to replace.

Tick the box to show which variable you would choose.

x ☐ y ☐

Follow this working and fill in the gaps.

Rearrange equation (1) to get $\qquad$ $y =$

Substitute into equation (2) $\qquad$ $= 3$

Multiply out

LINKS
See solving quadratics on p. 5.

Collect like terms and make $= 0$ $\qquad$ $2x^2 - 5x - 3 = 0$

Factorize

Solve to obtain $\qquad$ $x = -0.5$ and $x = 3$

Substitute $x = 3$ into (1) to give $\qquad$ $y =$

Substitute $x = -0.5$ into (1) to give $\qquad$ $y =$

So the coordinates for the points of intersection are

Turn the page for some exam questions on this topic ➤

Exam question 1

● ● ●

The line $y = 2x - 6$ meets the *x*-axis at the point *A*. The line $2y = x + 6$ meets the *y*-axis at the point *B*. The two lines meet each other at the point *C*. Show that triangle *ABC* is isosceles.

Find the coordinates of *A*, *B* and *C*.

DON'T FORGET
A sketch can often prove very useful.

Now show the triangle is isosceles.

LINKS
See coordinate geometry on p. 13.

EXAMINER'S SECRETS
Explain every step; you may get method marks even if you go wrong.

Exam question 2

● ● ●

Find the coordinates of the points of intersection of the line $x - y = 3$ with the curve $x^2 - 3xy + y^2 + 19 = 0$.

Algebra – inequalities

Inequalities can be solved using the same methods as for equations, but the answer is a range of values rather than a finite number of solutions.

LINEAR INEQUALITIES ○○○

The following inequalities have been solved incorrectly. Highlight where the mistake has been made and write underneath what has gone wrong.

$2x + 1 \leq 5$

$2x \leq 5 - 1$

$2x \leq 4$

$x < 2$

$8 - 5x > 23$

$-5x > 23 - 8$

$-5x > 15$

$x > -3$

$2x + 1 > 4x - 7$

$1 + 7 > 4x - 2x$

$8 > 2x$

$4 > x$

$x > 4$

DON'T FORGET
When multiplying or dividing by a negative number, change the inequality sign.

THE MODULUS SIGN ○○○

Remember that the modulus sign makes everything positive, so you need to create two separate inequalities showing the positive and negative options. To solve $|2x-1| \leq 9$ consider

$(2x - 1) \leq 9$ and $-(2x - 1) \leq 9$

Solve these two inequalities separately.

Now combine them to form one inequality.

QUADRATIC INEQUALITIES ○○○

EXAMINER'S SECRETS
A valid method is to sketch the graph then show its critical values and the required region.

Follow these steps every time and you shouldn't go wrong.

(1) Let quadratic = 0. (2) Solve quadratic to find critical values. (3) Sketch the graph. (4) Pick the region required. Try this one: $2x^2 + 9x - 5 \geq 0$.

Step 1 Let $2x^2 + 9x - 5 = 0$

Step 2 Factorize

Critical values

Step 3 Sketch the graph

Fill in the gaps.

WATCH OUT
We can combine $x \geq -5$ and $x \leq 0.5$ into $-5 \leq x \leq 0.5$. We cannot combine $x \leq -5$ and $x \geq 0.5$ so we leave them separate.

Step 4 We require the region where the graph ≥ 0, i.e. the x-values for which the graph is above (or on) the x-axis.

So the solution is

MATHS

Turn the page for some exam questions on this topic ➤

WATCH OUT
This inequality is $<$ not $\leq$ so be careful with your answer; don't include the $=$ part.

EXAM QUESTION 1

Use algebra to solve $(x-4)(x+5) = -8$. Hence or otherwise find the set of values of x for which $(x-4)(x+5) \subset -8$.

First solve the equality

Now solve the inequality

EXAM QUESTION 2

Sometimes inequalities can be used to solve practical problems. A farmer needs to fence a rectangular enclosure for their sheep. The length of the enclosure x is to be 8 m more than its width. The farmer has 80 m of fencing in stock, which is the maximum the farmer can use without buying some more. The farmer wants the area of the field to be greater than 240 m².

Start with a sketch showing dimensions.

Draw a sketch, then form a linear inequality about the perimeter and solve it for x.

Form a quadratic inequality in x and solve it.

Determine the set of possible values for x.

Now combine your answers to (a) and (b) and use some common sense.

Coordinate geometry – straight lines

The geometry of straight lines is simplified by a few standard formulae; understand the formulae and you're halfway there.

THE STANDARD FORMULAE ○○○

The gradient of a line is $m = \dfrac{y_2 - y_1}{x_2 - x_1}$

The gradient of a parallel line is m
The gradient of a perpendicular line is $-1/m$
There are three ways to write the equation of a line

Fill in the five formulae.

DON'T FORGET
You may want to use the points A (x_1, y_1) and $B (x_2, y_2)$ in your formulae.

1   2   3

Midpoint of a line segment is at

Distance between two points =

FINDING THE EQUATION OF A LINE ○○○

Find the equation of the line which passes through (6,7) and (3,3).

Step 1 Find the gradient using $m = \dfrac{y_2 - y_1}{x_2 - x_1} = \dfrac{7-3}{6-3} = \dfrac{4}{3}$

Step 2 Find the line's equation using $y = mx + c$ or $y - y_1 = m(x - x_1)$

Complete the working.

DON'T FORGET
You can use either point (6,7) or (3,3); the answer will end up the same.

EXAMINER'S SECRETS
It always looks impressive if you can get rid of any fractions from your equation, so multiply through by the denominator.

Find the equation of the line which passes through the midpoint of $A (-3, 4)$ and $B (6, -2)$ and is perpendicular to the line $2y - 3x + 5 = 0$.

Step 1 Find the midpoint of A and B using
$$\left(\frac{x_1 + x_2}{2}, \frac{y_1 + y_2}{2} \right) = \left(\frac{-3 + 6}{2}, \frac{4 - 2}{2} \right) = (\tfrac{3}{2}, 1)$$
Step 2 Rearrange $2y - 3x + 5 = 0$ to get it in the form $y = mx + c$

Complete the working.

Step 3 Find the required equation from the midpoint and gradient

Use your favourite method and rearrange to get in the form $ax + by + c = 0$.

PARALLEL AND PERPENDICULAR LINES ○○○

For some quick practice, tick the box to indicate whether the pairs of lines given are parallel, perpendicular or neither.

		Parallel	Perpendicular	Neither
$4y - 2x = 11$	$2y = x + 10$			
$3y - 4x = 7$	$3y + 4x = 1$			
$3y + 15 = x$	$y + 3x = 12$			

Turn the page for some exam questions on this topic ➤

EXAM QUESTION 1

● ● ●

The line L passes through the points shown. **(a) Calculate the distance between A and B. (b) Find an equation for L in the form $ax + by + c = 0$, where a, b and c are integers.**

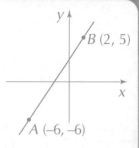

DON'T FORGET
All you need to find the equation of a line is its gradient and a point on the line.

EXAM QUESTION 2

● ● ●

EXAMINER'S SECRETS
Work your way methodically through the question one step at a time. Show all your working and quote all formulae before you use them. This shows the examiner you know the methods.

Here is part of the curve with equation $y = 10x - kx^2$; k is a constant. A is $(0,16)$. The straight line L, which passes through A, B and C, is parallel to the line with equation $3y + 6x = 5$.

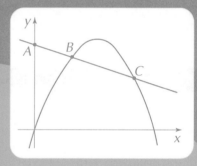

(a) Obtain an equation for L in the form $y = mx + c$. **(b)** Given the coordinates of C are $(4, 8)$, find the value of k. **(c)** Calculate the coordinates of B.

Take one step at a time. First find an equation for L.

Now use the coordinates you've been given and the equation of the curve to find the value of k.

B is one of the points where line L cuts the curve, so solve the curve equation and the line equation simultaneously.

LINKS
See simultaneous equations on p. 9.

Coordinate geometry – lines and curves

Sketching and plotting curves and straight lines is a very important topic, and examiners like to test that you can do it.

STRAIGHT LINES ○○○

Any straight line can be written in the form $y = mx + c$, where m is the gradient and c is the intercept on the y-axis.

> Draw these straight lines. First plot the intercept then use the gradient to plot two more points.

$2y = x - 4$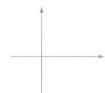

$y + x = 3$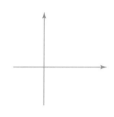

GRAPH SKETCHING ○○○

To sketch a curve only requires the basic shape to be known and the coordinates where the curve crosses the axes.

> Draw graphs of these basic functions; axes have been provided for you.

| $y = x$ | $y = x^2$ | $y = x^3$ | $y = 1/x$ |

GRAPH PLOTTING ○○○

To plot a curve requires points to be calculated and the curve accurately drawn on graph paper. Plot the curve with equation $y = x^3 + 2x^2 - 4$ for values of x from -3 to $+3$.

> Complete the table of values then plot the curve accurately on graph paper.

DON'T FORGET
Bigger scales mean greater accuracy.

SYLLABUS CHECK
The circle is required on the following modules: AQA(A)-P2, AQA(B)-P4, Edexcel, OCR-P3, MEI-P1, WJEC-P2.

x	y	x	y
-3		1	
-2		2	
-1		3	
0			

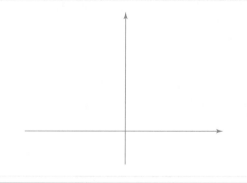

THE CIRCLE ○○○

Equation of a circle centre (a, b) with radius r

> Write the two standard formulae for the equation of a circle.

Equation of a circle centre at $(-g, -f)$ with $r^2 = g^2 + f^2 - c$

> Fill in the gaps and complete the working.

Find the equation of the circle centre $(6, 3)$ and radius 2; give your answer in the form $x^2 + y^2 + 2gx + 2fy + c = 0$

Use $(x - a)^2 + (y - b)^2 = r^2$

Multiply out

Then simplify

EXAMINER'S SECRETS
Get used to both formulae. If you're not told which one to use, then go for the one that involves less work.

Turn the page for some exam questions on this topic ➤

EXAM QUESTION 1

(a) Plot the graph of $y = \dfrac{2}{3-x}$ $(x \neq 3)$ for values of x from -1 to $+7$.

(b) On the same graph draw the line with equation $2y + x = 2$.

(c) Hence give the solutions of $1 = \dfrac{2}{3-x} + \dfrac{x}{2}$; express them to 1 d.p.

Set up a table of values before plotting the graph.

(a) Plot the graph

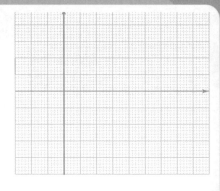

(b) Draw the line

Start by rearranging the equation of the line into the form $y = mx + c$.

(c) Solve the equation

Think about how you could rearrange the equation you've been asked to solve, and notice the question says 'hence'.

EXAM QUESTION 2

DON'T FORGET
Quote any formulae before you use them; it shows you know the methods.

(a) A circle C has the equation $x^2 + y^2 - 4x + 2y - 4 = 0$. Find its centre and radius by rearranging the equation in the form $(x - a)^2 + (y - b)^2 = r^2$.

(b) Draw circle C accurately on graph paper.

(c) The line $y = 2x$ intersects C. Add this line to your drawing in (b).

(d) Hence find the solutions to the simultaneous equations $y = 2x$ and $x^2 + y^2 - 4x + 2y - 4 = 0$; give you answers to 1 d.p.

Rearrange then complete the square.

(a)

(b), (c)

LINKS
See completing the square on p. 7.

Use your graph to find where the line intersects the circle.

(d)

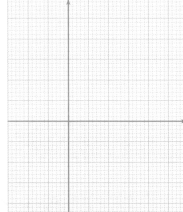

Trigonometry – radians

Angles are usually measured in degrees. A radian is a larger unit often used in trigonometry. You need to be confident working in degrees and radians.

DON'T FORGET
$360° = 2\pi^c = 2\pi$ rad.

DEGREES AND RADIANS ○○○

Remember that $30° = \pi/6$ rad; you can build a lot of useful angles from $30°$. For example, $150° = 5 \times 30° = 5\pi/6$ rad.

Complete the table by filling in the appropriate radian or degree equivalent. Leave the radian answers in terms of π.

Deg		90		60		30		15
Rad	2π		3π		π		$\frac{1}{4}\pi$	

Deg		210		150		720		120
Rad	$\frac{3}{2}\pi$		$\frac{11}{6}\pi$		$\frac{5}{3}\pi$		$\frac{5}{2}\pi$	

There is no need to avoid adding and subtracting in radians. Have a look at this example. Then complete the others.

DON'T FORGET
When adding or subtracting radians, take out π, treat as normal fractions then put back π at the end.

$$\frac{\pi}{6} + \frac{\pi}{2} = \pi(\tfrac{1}{6} + \tfrac{1}{2}) = \pi(\tfrac{1}{6} + \tfrac{3}{6}) = 2\pi/3$$

$$\frac{\pi}{3} + \frac{\pi}{4} =$$

$$\frac{\pi}{4} - \frac{\pi}{6} =$$

$$\frac{2\pi}{3} + \frac{\pi}{2} =$$

$$2\pi - \frac{\pi}{6} =$$

$$\frac{3\pi}{4} + \frac{3\pi}{2} =$$

CIRCULAR MEASURE ○○○

Complete these three formulae for θ measured in radians.

Arc length   =

Area of sector   =

Area of a triangle =

The diagram shows a sector of a circle.

(a) Find the arc length
(b) Find the area of the sector (2 d.p.)
(c) Find the area of the triangle (2 d.p.)
(d) Hence find the shaded area

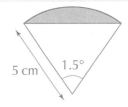

5 cm   1.5°

Complete the working and give units for your answers.

(a) Arc length $r\theta =$

(b) Sector area   =

(c) Triangle area =

DON'T FORGET
When using $\frac{1}{2}ab \sin C$, if your angle's in radians, make sure your calculator is in radian mode.

(d) Shaded area =

Turn the page for some exam questions on this topic ➤

Exam question 1

● ● ●

There is a straight path of length 50 m from point *A* to point *B* on a river bank. The river forms an arc of a circle centre *C*, radius 32 m.

(a) Calculate to 2 d.p. the size, in radians, of angle *ACB*.
(b) Calculate to 2 d.p. the length of this bend in the river.
(c) Calculate to 2 d.p. the shortest distance from *C* to the path.
(d) Calculate to 3 s.f. the area enclosed by the path and the river.

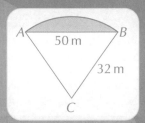

DON'T FORGET
Circular measure may involve basic trig, so recall **SohCahToa**. Sine is opposite over hypotenuse. Cosine is adjacent over hypotenuse. Tangent is opposite over adjacent.

First find angle *ACB*; think about an easy way to do it. Store the exact value in your calculator memory and use it in all further calculations. It makes your answers as accurate as possible.

(a)

Now find the length of the river bend.

(b)

Next find the shortest distance.

(c)

(d)

Exam question 2

● ● ●

Here is a circle centre *O* and radius *r*, with chord *XY*. Angle *XOY* is 2α radians. In terms of *r* and α, find (a) the length of the minor arc *XY* and (b) the length of the chord *XY*. (c) If the length of the minor arc *XY* is $1\frac{1}{2}$ times the length of the chord *XY*, show that $2\alpha - 3 \sin \alpha = 0$.

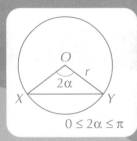

$0 \le 2\alpha \le \pi$

Hint: split triangle *OXY* into two right-angled triangles.

(a) Length of minor arc *XY*

EXAMINER'S SECRETS
If you're given a 'show that' question, make sure you show enough working to prove you've done it and not just copied the answer; the examiner won't be fooled.

(b) Length of chord *XY*

(c) Show that $2\alpha - 3 \sin \alpha = 0$

Trigonometry – graphs

Graphs of trig functions help to solve trig equations. You need to know their basic shapes, their maximum and minimum values, and how often they repeat themselves (i.e. their period).

GRAPHS OF SIN x, COS x AND TAN x ○○○

On the axes provided, sketch the graphs for sin x, cos x and tan x on the interval $0° \le x \le 360°$.

$y = \sin x$ 　　　　　 $y = \cos x$ 　　　　　 $y = \tan x$

Fill in the gaps giving the maximum and minimum values (where possible) and the period of each graph.

max =　　min =　　　　max =　　min =

period =　　　　　　　period =　　　　　　period =

GRAPH SKETCHING   ○○○

Quite often you'll be asked to sketch a graph. All you need is the basic shape and the points where the graph crosses the axes.

Use your standard graphs above to help sketch these. Describe in words how each standard graph is transformed.

$y = \sin(x - 30°)$ 　　 $y = 2\cos x$ 　　　 $y = \tan \frac{1}{2}x$
$0° \le x \le 360°$ 　　　 $0° \le x \le 360°$ 　　 $-180° \le x \le 180°$

LINKS
See transformations of graphs on p. 45.

GRAPH PLOTTING   ○○○

Accurately calculate several points then plot them on graph paper. Join the points with a smooth curve. Plot the curve $y = 1 + \sin 2x$ for $0° \le x \le 180°$.

Complete the table of values for y then plot the points on graph paper and join them with a nice smooth curve.

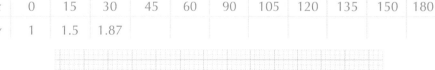

x	0	15	30	45	60	90	105	120	135	150	180
y	1	1.5	1.87								

Add the line $y = 1.6$ to your graph, then read off the solutions, i.e. the x-values where the curve and the line intersect.

DON'T FORGET
The bigger the scale, the more accurate the answer.

Use your graph to find approximate solutions to $1 + \sin 2x = 1.6$.

Turn the page for some exam questions on this topic ➤

EXAM QUESTION 1 ● ● ●

The diagram shows part of the curve with equation $y = A + B \sin 3x$ where A and B are constants and x is in degrees. The curve passes through the points $(0°, 1)$ and $(30°, 3)$ and meets the x-axis at the points L, M and N.

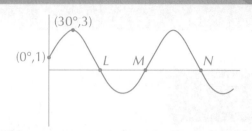

(a) Find the values of A and B.
(b) Hence determine the x-coordinates of L, M and N.

You've been told the coordinates of two points on the curve. Use them to write some equations involving A and B.

(a) Find A and B

(b) Find the x-coordinates for L, M, N

Remember that you need three solutions.

EXAM QUESTION 2 ● ● ●

(a) Sketch a graph of $y = 3 \cos \frac{1}{2}x$ for $0° \leq x \leq 720°$. State the maximum and minimum values of y and the period of the graph.
(b) Sketch a graph of $y = 2 \tan(x + 60°)$ for $-180° \leq x \leq 180°$; state the period.

Remember to mark on the points where the graphs cross the axes.

(a) $y - 3 \cos \frac{1}{2}x$

(b) $y = 2 \tan(x + 60°)$

DON'T FORGET
With any tan x graph you need to indicate the asymptotes (the dashed lines that the curve cannot cross).

Trigonometry – solving equations

Trigonometry is a topic you first met at GCSE; it becomes very important in pure maths and in applied maths.

SPECIAL ANGLES   ○○○

It's very useful if you can remember the ratios for particular angles. Complete the table then learn it.

Angle θ (degrees)	Angle θ (radians)	$\sin \theta$	$\cos \theta$	$\tan \theta$
0				
30				
45				
60				
90				
180				

SOLVING TRIGONOMETRIC EQUATIONS   ○○○

These formulae will help you to solve trigonometric (trig) equations:

$$\sin^2 x + \cos^2 x = 1 \qquad\qquad \tan x = \frac{\sin x}{\cos x}$$

Solve $\cos(x + 30) = \frac{1}{2}$ $(0° \leq x \leq 360°)$

Finish off these equations.

$$\cos^{-1}\left(\tfrac{1}{2}\right) =$$

$$\therefore x + 30 =$$

$$\therefore x =$$

Solve $\sin^2 \theta + \cos \theta + 1 = 0$ $(-180° \leq \theta \leq 180°)$
From $\sin^2 \theta + \cos^2 \theta = 1$ we have $\sin^2 \theta = 1 - \cos^2 \theta$
Substitute for $\sin^2 \theta$ in the equation

Solve $\sin \theta - \tan \theta = 0$ $(-\pi \leq \theta \leq \pi)$

Using $\tan \theta = \dfrac{\sin \theta}{\cos \theta}$ we have $\sin \theta - \dfrac{\sin \theta}{\cos \theta} = 0$

Factor out $\sin \theta$ in the equation

DON'T FORGET
Use a CAST (cosine, all, sine, tangent) diagram or a graph to find all the solutions.

EXAMINER'S SECRETS
Look carefully at the range of values you're given; it tells you whether your answers should be in degrees or radians.

DON'T FORGET
Solving an equation like $\sin(x + 60°) = 0.5$, list all solutions for $\sin^{-1} 0.5$ then subtract 60° from each one.

DON'T FORGET
Solving an equation such as $\tan 2\theta = 1$, you'll need to list solutions outside the range to start off with, because you'll be halving them all at the end.

Turn the page for some exam questions on this topic ➤

EXAM QUESTION 1

● ● ●

(a) Find the coordinates of the point where $y = 2\sin(2x + \pi/6)$ crosses the y-axis. (b) Find the values of x, $0 \le x \le 2\pi$, for which $y = \sqrt{2}$.

Start by asking yourself what is special about the x-coordinate of a graph when it crosses the y-axis.

DON'T FORGET
$45° = \pi/4$ rad $180° = \pi$ rad
$90° = \pi/2$ rad $360° = 2\pi$ rad

LINKS
See rationalizing surds on p. 39

EXAM QUESTION 2

● ● ●

Given that $\tan 15° = 2 - \sqrt{3}$, find, in the form $p + q\sqrt{3}$, where p and q are integers, the values of $\tan 75°$ and $\tan 165°$.

Draw a right-angled triangle with one angle of 15°. What's the size of the third angle?

EXAM QUESTION 3

● ● ●

Find all angles in the range 0–360° which satisfy $2\tan x = 1 + \dfrac{3}{\tan x}$.

Polynomials – remainder theorem

Make sure you're confident about multiplying out brackets and factorizing quadratics and cubics. Practise on these items.

LINKS
See expanding and factorizing on p. 5.

Expand these brackets then simplify.

EXPANDING BRACKETS ○○○

$x(x + 2)(x + 3) =$

$(x - 1)^2(x + 4) =$

$(2x - 3)^3 \quad =$

DIVIDING POLYNOMIALS BY LONG DIVISION ○○○

Using long division, divide $x^3 + 2x^2 - x - 2$ by $(x - 1)$

Now find the remainder when $3x^3 - 4x - 3$ is divided by $x^2 + 2x + 1$

Be careful, $3x^3 - 4x - 3$ has no x^2 term, so what do you do?

SYLLABUS CHECK
Check your syllabus to see whether you need to know the remainder theorem.

Find the remainder when the polynomials are divided by the linear expressions given.

REMAINDER THEOREM: A REMINDER ○○○

If polynomial $f(x)$ is divided by $(x - a)$ and the remainder is a constant, then the remainder is equal to $f(a)$. If $f(x) = x^2 + 3x - 4$ is divided by $(x - 4)$ then the remainder $= f(4) = 4^2 + (3 \times 4) - 4 = 24$.

$x^3 + 2x^2 - x + 5$ divided by $(x - 1)$

$2x^3 - 5x + 16$ divided by $(x + 2)$

$x^3 - 4x^2 - 7$ divided by $(2x - 1)$

$9x^3 + 3$ divided by $(3x + 2)$

Turn the page for some exam questions on this topic ➤

EXAM QUESTION 1 ● ● ●

(a) Expand and simplify $(x-1)(2x+3)(4-x)$ arranging your answer in descending powers of x.

(b) Find the remainder when the answer to (a) is divided by $x^2 + x + 1$.

EXAMINER'S SECRETS
Double-check this as you're going to use it in the next part – it's worth the effort.

(a) First expand and simplify

(b) Now use long division

EXAM QUESTION 2 ● ● ●

The remainder obtained when $3x^3 - 6x^2 + ax - 1$ is divided by $x+1$ is equal to the remainder when the same expression is divided by $x-3$. Find the value of a.

EXAM QUESTION 3 ● ● ●

WATCH OUT
Look closely at a question to see if you're given any conditions for your unknown number. In this question you're told $a \neq 0$.

The expression $6x^2 + x + 7$ leaves exactly the same remainder when it is divided by $x-a$ and $x+2a$ $(a \neq 0)$. Calculate the value of a.

Polynomials – factor theorem

Following on from the remainder theorem comes the factor theorem. We can use it to find roots of polynomials and hence sketch their graphs.

FACTOR THEOREM ○○○

If $(x - a)$ is a factor of a polynomial $f(x)$ then $f(a) = 0$ (i.e. there is no remainder). So to factorize the cubic $f(x) = x^3 - 2x^2 - 5x + 6$ we follow these three steps.

Step 1 Use the remainder theorem, and trial and improvement, to find a factor. Notice that the constant at the end is a 6, so it's worth trying a factor of 6, e.g. 1, 2 or 3. Try $x = 1$ so that $f(1)$ is

> Do the working and state the factor.

Step 2 Use long division, or a similar method, to divide the cubic by its factor and find the quadratic.

> Show your method here.

Step 3 Factorize the quadratic to give you the other two factors then write the factorized expression for $f(x)$.

> Factorize the quadratic to find the other two factors; then write $f(x)$ in factors.

SKETCHING GRAPHS OF POLYNOMIALS ○○○

To sketch the graph of a polynomial, first use the factor theorem to find the factors. Then, by equating the polynomial to zero, find its roots and mark them on the sketch. Finally decide in between two of the roots whether $f(x)$ is positive or negative. Now you're ready to sketch the graph.

If $f(x) = 2x^3 + 7x^2 - 7x - 12$ factorizes to $f(x) = (2x - 3)(x + 4)(x + 1)$ then solving $f(x) = 0$ gives $(2x - 3)(x + 4)(x + 1) = 0$; $x = \frac{3}{2}, -4, -1$.

> Mark $f(0)$ on the axis then sketch the curve.

DON'T FORGET
Finding $f(0)$ is a good idea anyway, as on your sketch you need to mark the point where the graph crosses the y-axis as well as the x-axis.

Choose a value between two of the roots, e.g. $x = 0$
Here $f(0) = -12$

Turn the page for some exam questions on this topic ➤

EXAM QUESTION 1

● ● ●

(a) Show that $2x + 1$ is a factor of $f(x) = 2x^3 + x^2 - 8x - 4$.

(b) Hence find the values of x for which $f(x) = 0$.

DON'T FORGET

To factorize $x^2 - y^2$ use the difference of two squares, so $x^2 - y^2 = (x + y)(x - y)$.

EXAMINER'S SECRETS

The phrase 'find the exact value' means your answer is either a whole number or it must be left as a fraction or surd. This is a big hint for part (b).

LINKS

See solving quadratics on p. 5; see trigonometric equations on p. 21.

EXAM QUESTION 2

● ● ●

Given that $f(x) = x^3 + x^2 - 5x - 2$, (a) show that $(x - 2)$ is a factor of $f(x)$;

(b) factorize $f(x)$ completely and find the exact values for which $f(x) = 0$; (c) hence, by substituting $x = \tan t$, solve the equation $\tan^3 t + \tan^2 t - 5 \tan t - 2 = 0$ ($0 \le t \le 360°$); give the solutions to 1 d.p.

Polynomials – binomial theorem

Expanding a binomial is relatively easy as long as you can remember either Pascal's triangle (if the power is small) or the expansion formula (if the power is large).

PASCAL'S TRIANGLE ○○○

Complete Pascal's triangle to show the coefficients when $(a + b)$ is expanded up to the power of 6.

$(a + b)^1$ 1   1

$(a + b)^2$ 1   2   1

$(a + b)^3$

$(a + b)^4$

$(a + b)^5$

$(a + b)^6$

First expand $(a + b)^5$.

Use Pascal's triangle to expand $(2 - x)^5$ in increasing powers of x.

Write out the expansion of $(2 - x)^5$ and simplify each term as far as possible.

Now compare $(a + b)^5$ with $(2 - x)^5$ and substitute $a = 2$ and $b = -x$.

BINOMIAL THEOREM ○○○

There are two versions of the binomial expansion; practise both.

Complete the expansions for $(a + x)^n$ up to the term in x^3.

$(a + x)^n = a^n + {}^nC_1a^{n-1}x +$

DON'T FORGET
$n! = n(n-1)(n-3)\ldots3.2.1$

$(a + x)^n = a^n + n.a^{n-1}x +$

$${}^nC_r = (\frac{n}{r}) = \frac{n!}{r!(n-r)!}$$

Use the first expansion and simplify your answer.

Expand $(1 + 2x)^{12}$ in ascending powers of x up to and including the term in x^4. Hence evaluate $(1.02)^{12}$ correct to 5 d.p.

Substitute $x = 0.01$ into the expansion then evaluate your answer.

Comparing $(1.02)^{12}$ with $(1 + 2x)^{12}$ then $1.02 = 1 + 2x$ and $x = 0.01$

Write the expansion up to the x^2 term.

The coefficient of x^2 in the expansion of $(1 + \frac{1}{2}x)^n$ is 7. Find n.

Simplify and solve to find n.

We're told the coefficient of x^2 is 7, so $7 = n(n-1)(\frac{1}{2})^2/2!$

MATHS

Turn the page for some exam questions on this topic ➤

Exam question 1

● ● ●

(a) Use the binomial theorem to expand $(2 + 10x)^4$, giving each coefficient as an integer. (b) Use your expansion to find the exact value of $(1002)^4$ stating the value of x which you have used.

Exam question 2

● ● ●

EXAMINER'S SECRETS
Follow the instructions; give ascending powers of x when asked for ascending powers and descending powers when asked for descending powers.

(a) Find the non-zero values of a and b in the expansion of $(a + x/b)^8$ in ascending powers of x, when the first term is 256 and the coefficient of x^2 is three times the coefficient of x^3.

(b) Using your values of a and b, give the first four terms in the binomial expansion of $(a + x/b)^8$ in descending powers of x.

Differentiation

Differentiation is for calculating the gradient of the tangent to a curve at any point on the curve. If $y = f(x)$ then we can write dy/dx as $f'(x)$ or simply y'; they all mean the same thing.

WHAT YOU SHOULD KNOW ○○○

Highlight the correct answer – A, B or C.

If the gradient of a tangent to a curve is m, then the gradient of the normal is what?
(A) m^2 (B) $1/m$ (C) $-1/m$

To find a maximum or minimum point on the curve we find dy/dx and set it equal to what?
(A) x (B) 0 (C) 1

To find the values of x for which $f(x)$ is an increasing function, state the condition we solve for $f'(x)$.   (A) $f'(x) = 0$ (B) $f'(x) < 0$ (C) $f'(x) > 0$

To find the values of x for which $f(x)$ is a decreasing function, state the condition we solve for $f'(x)$.   (A) $f'(x) = 0$ (B) $f'(x) < 0$ (C) $f'(x) > 0$

What is the result of differentiating $y = kx^n$?
(A) $dy/dx = knx^{n-1}$ (B) $dy/dx = \frac{k}{n}x^{n-1}$ (C) $dy/dx = kx^{n+1}/(n+1)$

USE OF DIFFERENTIATION WITH CURVE SKETCHING ○○○

Differentiating to find maxima and minima can help when sketching graphs of functions. Sketch the graph of $y = 2x^3 + 3x^2 - 12x + 1$.

Step 1 Find, when possible, the coordinates where the curve crosses the axes.

Put $x = 0$ then $y = 0$ into the equation of the curve to find where it crosses the axes.

Step 2 Find any turning points by differentiating.

Differentiate then put $dy/dx = 0$; solve for x.

DON'T FORGET
Once you've found the x-coordinates of any turning points, find their corresponding y-values.

Step 3 Decide whether the turning points are maxima, minima or points of inflexion.

Examine the sign of dy/dx before and after the turning point to determine its nature.

Finally sketch the curve.

Turn the page for some exam questions on this topic ➤

EXAM QUESTION 1

(a) Find the coordinates of any stationary points on the curve $f(x) = 5x^6 - 12x^5$ and distinguish between them.
(b) Hence sketch the curve and give the values of x for which $f(x)$ is an increasing function.

DON'T FORGET
$f'(x) > 0$ for an increasing function.

EXAM QUESTION 2

Differentiation is often used for finding the answers to practical problems.

An open metal tank with a square base is made from 16 m² of sheet metal. Find the tank dimensions needed to maximize the volume and find this maximum volume.

Start with a sketch showing the letters you're using for the dimensions of the tank, then use the information you're given: area is 16 m² and the volume is to be found.

DON'T FORGET
Check it's a maximum by examining the sign of dy/dx.

Integration

Integration is the reverse of differentiation – given the gradient function of a curve (and a point on the curve) we can integrate to obtain the equation of the curve.

THE REVERSE OF DIFFERENTIATION ○○○

Link each gradient function with the correct curve equation; use the values for x and y to work out the constant.

DON'T FORGET

If $\dfrac{dy}{dx} = kx^n$ then

$y = \dfrac{kx^{n+1}}{n+1} + c \ (n \neq -1)$.

1 $dy/dx = 2x + 5$	$x = 1, y = 0$	$y = \frac{1}{3}x^3 - 2x + 2$
2 $dy/dx = x^2 - 2$	$x = 3, y = 5$	$y = 2\sqrt{x} + 3x - 6$
3 $dy/dx = x^4 - 3x$	$x = 0, y = 2$	$y = x^2 + 5x + 1$
4 $dy/dx = 2x + 5$	$x = 0, y = 1$	$y = x^2 + 5x - 6$
5 $dy/dx = 1/\sqrt{x} + 3$	$x = 4, y = 10$	$y = \frac{2}{3}\sqrt{x^3} + x^2 - 11\frac{1}{3}$
6 $dy/dx = \sqrt{x} + 2x$	$x = 4, y = 10$	$y = \frac{1}{5}x^5 - \frac{3}{2}x^2 + 2$

THE AREA UNDER A CURVE ○○○

The area under a curve can be found using integration. If we require the area between a curve and the x-axis, we use $\int y \, dx$ and if we require the area between a curve and the y-axis we use $\int x \, dy$. To find the area in the first quadrant between the curve $y^2 = x$, the y-axis and the line $y = 3$ follow these steps.

Step 1 Find where the curve intersects the axes then sketch the curve, add the line $y = 3$ and shade the required area.

Complete step 1.

Step 2 Decide whether $\int y \, dx$ or $\int x \, dy$ is needed.

Complete step 2.

Step 3 Integrate and substitute the limits.

Complete step 3.

Area $= \left[\frac{1}{3}y^3\right]_0^3 =$

Follow the same three steps to find the area between the curve $y = 2x - x^2$ and the x-axis from $x = 0$ to $x = 3$.

Complete the three steps again, but think carefully how to find the area required.

Step 1

EXAMINER'S SECRETS
Always show you've substituted in both limits, even if you can see that one is zero. It tells the examiner you know what you're doing.

Step 2

DON'T FORGET
The area lying above the axis will have a positive value and the area lying below will have a negative value, so be careful.

Step 3

The negative sign on area *B* indicates that it lies below the x-axis; remove the negative sign and add to area *A*.

Turn the page for some exam questions on this topic ➤

EXAM QUESTION 1

Find the area bounded by $y^2 = x + 4$, the y-axis and the line $y = 4$.

EXAM QUESTION 2

Find the area of the region enclosed between the line $y = 2$ and the curve $y = 5x - 4 - x^2$.

First find where the line and curve intersect then do a sketch.

WATCH OUT
Don't try to do too much in your head – write it all down. It might take longer but you're less likely to make a mistake.

EXAM QUESTION 3

Find the area bounded by the curves $y = x^2 - 4x$ and $y = 6x - x^2$.

Find where the two curves intersect then do a sketch.

Integration – trapezium rule

The trapezium can be used to find an approximate value for the area under a curve, without having to integrate.

USING THE TRAPEZIUM RULE ○○○

If the area is divided into n strips, each of width h, then the formula is

Fill in the formula for the trapezium rule.

DON'T FORGET
Five strips means there are six ordinates, i.e. there are six values for x and y.

Complete the table of values for y.

Complete the calculation.

Find an approximate value for $\int_0^1 (1 + x)^{-1}$ with five strips.

Since x goes from 0 to 1, then with five strips h would be $\frac{1}{5}$.

x	0	$\frac{1}{5}$	$\frac{2}{5}$	$\frac{3}{5}$	$\frac{4}{5}$	1
y	1					

$\therefore \int_0^1 (1 + x)^{-1} dx =$

ACCURACY OF THE TRAPEZIUM RULE ○○○

The more strips or ordinates used, the more accurate the answer.

If $A = \int_1^2 \dfrac{3x^4 + 2}{x^2} \, dx$

DON'T FORGET
Six ordinates means 5 strips, and 11 ordinates means 10 strips.

(a) Find A by integration. (b) Use the trapezium rule to find an approximate value for A by (i) taking 6 ordinates and (ii) taking 11 ordinates. (c) Calculate the percentage error in using the trapezium rule, giving your answer correct to 2 d.p.

Integrate this by separating it into two fractions and simplifying.

x goes from 1 to 2 so $h = 0.2$ (6 ordinates) and $h = 0.1$ (11 ordinates).

Complete the tables, giving the y-values correct to 3 d.p. then substitute into the trapezium rule and evaluate.

x	1	1.2	1.4			
y	5	5.709				

EXAMINER'S SECRETS
Use the memory on your calculator to store the accurate values (not the rounded values) then your final answer will be as accurate as possible.

x	1	1.1	1.2	1.3	1.4	1.5
y						

x	1.6	1.7	1.8	1.9	2	
y						

DON'T FORGET
$\% \text{ error} = \dfrac{\text{actual error}}{\text{actual value}} \times 100\%$

Calculate the percentage error.

% error $= 100 \times (8.031\ 5662 - 8)/8$ $=$

% error $=$ $=$

Turn the page for some exam questions on this topic ➤

EXAM QUESTION 1

Find an approximate value for $\int_0^{\pi/2} \sin x \, dx$ by using the trapezium rule with eight strips. Give your answer correct to 5 d.p.

In your table write your *y*-values correct to 3 d.p. but for greatest accuracy remember to use exact values in your calculations.

DON'T FORGET
When calculating the *y*-values make sure your calculator is in radian mode.

EXAM QUESTION 2

The diagram shows part of the curve $y = Ax^3 + Bx^2 + Cx + D$. (a) Find the values A, B, C, D. (b) Use the trapezium rule (5 ordinates) to estimate the value of the shaded area to 5 d.p. (c) Use algebraic integration to find the exact area of the shaded region. (d) Calculate the percentage error in using the trapezium rule to estimate this area, and say whether this estimate is an under- or overestimate.

LINKS
See the factor theorem on p. 25.

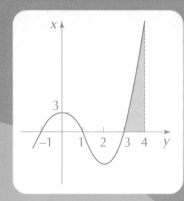

Integration – volumes of revolution

When the area under a curve is rotated 360° about an axis, the volume of the solid of revolution can be found by integration.

THE FORMULAE   ○○○

Fill in the two formulae you need to know.

If the curve is rotated about the x-axis then $V =$

If the curve is rotated about the y-axis then $V =$

ROTATION ABOUT THE X-AXIS   ○○○

Find the volume of the solid formed when the area bounded by the curve $y = 2x^2 + 3x$ is rotated through 360° about the x-axis between the ordinates $x = 1$ and $x = 2$; leave your answer in terms of π.

Complete step 1.

Step 1 If possible, sketch the curve to see where it cuts the axes.

Complete step 2.

Step 2 Formula uses y^2 so find y^2 in terms of x. Have $y = 2x^2 + 3x$ so

DON'T FORGET
To find y^2, put $2x^2 + 3x$ in a bracket then square the bracket.

Step 3 Now substitute y^2 into $V = \int \pi y^2 dx$ and integrate.
$V = \pi \int_1^2 (4x^4 + 12x^3 + 9x^2)\, dx = \pi \left[\frac{4}{5}x^5 + 3x^4 + 3x^3 \right]_1^2$

Complete step 4.

Step 4 Substitute the limits and evaluate.

EXAMINER'S SECRETS
Try to work in fractions and leave π in your answer; it means your answer will be accurate and it looks good too.

ROTATION ABOUT THE Y-AXIS   ○○○

Find the volume of the solid formed when the area bounded by the curve $y = x^2 + 2$ (in the first quadrant) and the line $y = 3$ is rotated through 360° about the y-axis.

Complete step 1.

Step 1 Draw a sketch of the curve if possible.

Complete step 2.

Step 2 Formula uses x^2 so find x^2 in terms of y. Have $y = x^2 + 2$ so

Complete step 3.

Step 3 Now substitute x^2 into $V = \int \pi x^2 dy$ and integrate.

Complete step 4.

Step 4 Substitute the limits and evaluate.

Turn the page for some exam questions on this topic ➤

EXAM QUESTION 1

Find the volume of the solid formed when the area enclosed by the curve $y = x - 1/x$, the x-axis and the line $x = 2$ is rotated through 360° about the x-axis.

DON'T FORGET

If the question asks for the answer correct to 2 d.p. or 3 s.f. then evaluate your final answer, otherwise leave π as it is.

EXAM QUESTION 2

(a) Sketch the curve $y = 2x^2 + 1$ and mark any points where the curve crosses the axes. (b) Find the area enclosed by the curve, the y-axis, the x-axis and the line $x = 3$. (c) Find the volume of the solid formed when the area in the first quadrant bounded by the curve, the y-axis and the line $y = 3$ is rotated through 360° about the y-axis.

EXAMINER'S SECRETS

Even if you're not asked for a sketch, it's often a good idea to draw something. A sketch can help you see how to approach the question and choose the limits for the integration.

Calculate the area.

Now calculate the volume.

Proof

When trying to prove something, make sure your working is clear, and check each step is logical and correct.

NOTATION USED ○○○

> Match each word statement with its corresponding symbol statement.

1 P is equal to Q	$P \approx Q$
2 P is implied by Q	$P \neq Q$
3 P is approximately equal to Q	$\sim Q$
4 Not Q	$P \equiv Q$
5 P implies Q	$P \nRightarrow Q$
6 P is not equal to Q	$P \Rightarrow Q$
7 P does not imply Q	$P \Leftarrow Q$
8 P is identical to Q	$P = Q$

DIRECT PROOF ○○○

Direct proof starts with facts which are accepted, then argues logically to the result which is required.

> See if you can finish the proof.

Prove that if a is odd and b is even then ab is even. Start by defining a and b in terms of another letter to do with a being odd and b being even. If a is odd it can be written as $a = 2p + 1$. If b is even it can be written as $b = 2q$. Substituting into ab we have

PROOF BY COUNTEREXAMPLE ○○○

To prove a statement is false, all we have to do is produce just one case which doesn't fit the statement.

> Find a counterexample for each of these.

$x^2 = y^2 \Rightarrow x = y$

DON'T FORGET
You just need to find one example which *doesn't* work.

For any two real numbers a, b we have $a/b = b/a$

For any two real numbers a, b we have $a - b > 0 \Rightarrow a^2 - b^2 > 0$

PROOF BY CONTRADICTION ○○○

Start by negating the statement you want to prove then show this negation is false; this then means the statement must be true.

SYLLABUS CHECK
Proof by counterexample and contradiction are required on the following modules: MEI-P1, AQA(A)-P2, Edexcel-P2, OCR-P3, WJEC-P2, AQA(B)-P4.

Prove by contradiction that if $f(x) = x^2 + bx + c$ then $b^2 - 4c < 0 \Rightarrow f(x) > 0$ for all real values of x.
Start by negating the statement, i.e. $b^2 - 4c < 0 \Rightarrow f(x) \leq 0$ for at least one real value of x.

> Now try to prove that the negated statement is false.

Turn the page for some exam questions on this topic ➤

For more on this topic, see pages 34–35 of the *Revision Express A-level Study Guide*

EXAM QUESTION 1

● ● ●

Prove that if a and b are both odd numbers, then $a + b$ is even.

Start by defining a and b in terms of other letters.

EXAM QUESTION 2

● ● ●

Prove the identity $(\sin x + \cos x)^2 + (\sin x - \cos x)^2 = 2$.

Start with one side then prove it's equal to the other.

LINKS

See trigonometric equations on p. 21.

EXAM QUESTION 3

● ● ●

Prove by contradiction that $x + 1/x \geq 2$ for all $x > 0$.

LINKS

See inequalities on p. 11.

Algebra – surds, indices and laws of logs

You need to know how to use indices, surds and logarithms.

INDICES ○○○

Draw a line to connect the expression on the left with its corresponding expression on the right.

1	$a^m a^n$		a^{mn}
2	a^m / a^n		$\sqrt[n]{a}$
3	$(a^m)^n$		1
4	a^0		$1/a^n$
5	a^{-n}		a^{m+n}
6	$a^{1/n}$		a^{m-n}

Using the laws of indices and without a calculator, find the exact value of $2^5 \times 8^3$.

You need to rewrite 8 as a power of 2.

SURDS ○○○

Surds are used to write down irrational numbers exactly.
You need to be able to simplify surds.
Simplify $\sqrt{32}$ and $(2 + \sqrt{2})(3 - \sqrt{8})$

THE JARGON
When asked to 'simplify' a surd, you need to write it with the smallest possible integer under the square root sign. You are looking to rewrite the surd in the form $\sqrt{x^2 y}$ which can be simplified to $x\sqrt{y}$.

You need to be able to rationalize the denominator.

THE JARGON
Rationalizing the denominator means eliminating any surds in the denominator.

Simplify $\dfrac{\sqrt{3}}{\sqrt{2}}$ by multiplying by $\dfrac{\sqrt{2}}{\sqrt{2}}$

LAWS OF LOGARITHMS ○○○

Write down the three laws of logarithms.

The power rule

The multiplication rule

The division rule

Turn the page for some exam questions on this topic ➤

EXAM QUESTION 1

Express $\dfrac{8}{\sqrt{8}} - \dfrac{2}{\sqrt{2}}$ in the form $a\sqrt{b}$ where a and b are integers.

Take your time and don't rush. Rationalize the denominator for each expression then simplify the fractions and the surds to obtain the answer.

EXAM QUESTION 2

Solve the equation $2^x = 10^{5x-1}$; give your answer to 3 s.f.

Take logs of both sides then use the power rule to rewrite the equation with the unknowns in front of the logarithms. Now you can solve the equation in the normal way. You'll need to use your calculator to evaluate any logarithms in your answer.

EXAM QUESTION 3

The strength of a particular radioactive source after t years is given by $R = 8000 \times 5^{-0.003t}$. State the initial value of R and find the value of t when the source has decayed to half its value; give your answer correct to three significant figures.

The initial value is found when $t = 0$.

Write an equation with R equal to half the initial value.

Take logs of both sides then solve.

Algebra – e^x and ln x

You need to be familiar with the exponential function and the related logarithmic function.

THE EXPONENTIAL FUNCTION e^x ○○○

The exponential function has a very important property.

> Write down the important property of e^x.

GRAPHING e^x AND ln x ○○○

> Sketch the graphs of e^x and ln x on the axes provided.

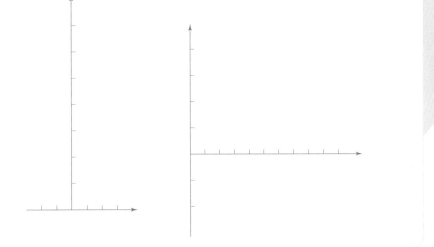

EXAMINER'S SECRETS
You are sketching a function and its inverse. This can be done by reflecting the function in the line $y = x$.

LINKS
See laws of logarithms on p. 39.

THE NATURAL LOGARITHM ln x ○○○

The inverse function of e^x is ln x; ln x obeys all the laws of logarithms.

> Find x when $e^{3x} = 7$.

> You'll need to use logarithms for this.

DIFFERENTIATING e^x AND ln x ○○○

Differentiating e^x is very easy. But what about differentiating ln x?

> Differentiate these functions.

$$\frac{d}{dx}e^{4x} =$$

$$\frac{d}{dx}(\ln x) =$$

Turn the page for some exam questions on this topic ➤

For more on this topic, see pages 24–25 of the *Revision Express A-level Study Guide*

EXAM QUESTION 1

● ● ●

In a lab experiment the relationship between the number of bacteria N present in a culture after time t hours is modelled by $N = 50e^{1.3t}$.

(a) When will the number of bacteria have reached 10 000? (b) What will be the rate of increase of bacteria per hour when $t = 8$? Give your answers to two significant figures.

(a) First find the time when $N = 10\,000$

(b) Now find the rate of increase dN/dt when $t = 8$

DON'T FORGET
Stationary points or turning points occur when the gradient is zero.

LINKS
See differentiation on p. 29.

EXAM QUESTION 2

● ● ●

The equation of a curve is $y = 4x^2 - 2\ln x$ where $x > 0$. Find the coordinates of the stationary point on the curve.

Sequences and series

You are studying lists of numbers constructed from a formula or an inductive definition.

SEQUENCE OR SERIES ○○○

Define sequence and series.

Sequence

Series

GENERATING SEQUENCES AND SERIES ○○○

You need to be able to generate a sequence from a formula or an inductive definition, and a series from its sigma notation.

Formula defintion

Write the first five terms defined by formula $u_n = 2n^2 - 1$.

Inductive defintion

Write the first five terms generated by $u_1 = 3$, $u_{n+1} = 3u_n + 2$.

Sigma defintion

Write out the series $\sum_{r=1}^{n} 2r^3$ up to its fifth term.

THE JARGON
Sigma is the Greek letter Σ. It means 'the sum of'.

ARITHMETIC AND GEOMETRIC PROGRESSIONS ○○○

An arithmetic progression (AP) has first term a and common difference d. The nth term is found using which formula?

Circle the correct formula.

$(n-1)d$ $\qquad$ $a + nd$ $\qquad$ $a + (n+1)d$ $\qquad$ $a + (n-1)d$

The sum of the first n terms S_n can be found using which formula?

Circle the correct formula.

$n(2a+d)$ $\qquad$ $\frac{1}{2}n(2a+nd)$ $\qquad$ $\frac{1}{2}n[2a+(n-1)d]$ $\qquad$ $\frac{1}{2}n[2a+(n+1)d]$

A geometric progression (GP) has first term a and common ratio r. The nth term is found by using which formula?

Circle the correct formula.

ar^{n-1} $\qquad$ ar^n $\qquad$ $a^{n-1}r$ $\qquad$ $a + (n-1)r$

The sum S_n of the first n terms can be found using which formula?

Circle the correct formula.

$a(1-r^n)$ $\qquad$ $\frac{a(1-r^n)}{1-r}$ $\qquad$ $\frac{a(1-r)}{1-r}$ $\qquad$ $\frac{1}{2}n[2a+(n-1)r]$

THE JARGON
Convergence is when a sequence or series tends to a limiting value.

Write down the condition for convergence.

Some geometric progressions converge and the sum to infinity is found by using the formula $a/(1-r)$. This can only be used when r satisfies a certain condition.

What is the sum of the numbers 1 to 100?

Turn the page for some exam questions on this topic ➤

EXAM QUESTION 1

● ● ●

The fourth term of an arithmetic progression is five times the first term, and the second term is 7. Find the first term, the common difference and the sum of the first ten terms.

> To answer this question, use the formula for the *n*th term of an AP hence set up two simultaneous equations in *a* and *d*.

EXAM QUESTION 2

● ● ●

A mortgage is taken out for £20 000 and is repaid by annual instalments of £4000. Interest is charged on the outstanding debt at 10% calculated annually.

If the first repayment is made one year after the mortgage is taken out, find the number of years it takes to repay the mortgage.

> Write expressions for the amount outstanding after 1, 2 and 3 years.

> Now write down the amount outstanding after *n* years.

> Simplify your expression and look for a GP.
> Rewrite your GP using the formula for the first *n* terms.

> After *n* years you want the outstanding amount to be zero. Use this to write an equation you can solve by logarithms.

LINKS
See logarithms on p. 39.

Functions

The idea behind functions such as $y = x^2$, $y = x^3 - 2$ and $y = 2x$ is that for every value of x, you can find a unique value of y.

DOMAIN AND RANGE ○○○

Define domain and range.

Domain

Range

Find the range of function $y = \sqrt{x(10 - x)}$ with domain $0 \leq x \leq 10$.

What does the root sign tell you about y?

Square both sides; form a quadratic in x.

By considering the discriminant write down an inequality for y and hence write down the range of the function.

LINKS
See quadratics on p. 5.

INVERSE FUNCTIONS ○○○

Find the inverse of the function $f(x) = \dfrac{2x + 3}{x}$, $x \neq 0$.

EXAMINER'S SECRETS
To find an inverse function first write the function as $y = f(x)$. Now make x the subject of the formula. Then swap x and y. You have found the inverse function.

TRANSFORMATIONS OF GRAPHS ○○○

If a curve is defined by the function $y = f(x)$, we can transform the graph of y by changing the expression involving $f(x)$.

Draw a line connecting the new $f(x)$ expression on the left to the resulting transformation on the right.

1	$f(x - a)$	Stretch in the y-direction, scale factor a
2	$f(x) + a$	Translation of a units in the x-direction
3	$f(ax)$	Translation of a units in the y-direction
4	$af(x)$	Stretch in the x-direction, scale factor $1/a$

Turn the page for some exam questions on this topic ➤

EXAM QUESTION 1

Sketch the function $y = x^2$. Hence sketch graphs of (a) $y = (x + 2)^2$ and (b) $y = 4x^2 + 1$. Show clearly where the graphs cross the axes.

First sketch the function $y = x^2$

(a) Now sketch $y = (x + 2)^2$

(b) And finally $y = 4x^2 + 1$

Think of each graph as a transformation of the original function then sketch it.

EXAM QUESTION 2

For what value of x is the function $f(x) = 3/(2 - x)$ undefined?

Find the inverse function $f^{-1}(x)$.

MATHS

Differentiation – further functions

Having mastered the basics, you need to apply them in finding equations of tangents and normals, and you need to know how to differentiate products and quotients.

PRODUCT RULE AND QUOTIENT RULE ○○○

These rules are easy to remember in terms of u and v, where u and v are both functions of x.

Write the product rule here.

The product rule tells you how to differentiate the product $y = uv$.

Write the quotient rule here.

The quotient rule tells you how to differentiate the quotient $y = u/v$.

LINKS
See e^x on p. 41.

Differentiate $y = e^x(4x^2 + 3)$ with respect to x.

Differentiate $y = (2x + 1)/(3x - 1)$ with respect to x.

CHAIN RULE ○○○

The chain rule is very useful for differentiating composite functions.

Find dy/dx when $y = (2 + x^3)^5$ by using the substitution $u = 2 + x^3$.

Begin by writing down the chain rule.

Substitute $u = 2 + x^3$ into the expression for y then apply the chain rule.

Turn the page for some exam questions on this topic ➤

EXAM QUESTION 1

Find the equation of the tangent to the curve $y = e^x(10 + 2x)$ when $x = -1$. Give your answers to 3 s.f.

To find the equation of a tangent you first have to find the gradient. For this you will have to differentiate by using the product rule. You will also have to find the value of y when $x = -1$. From this information you can now find the equation of the tangent.

EXAM QUESTION 2

WATCH OUT
This curve has an asymptote.

LINKS
See differentiation on p. 29.

Find the coordinates of the stationary point on the curve $y = (3 - x)/x^2$ where $x > 0$, and determine whether it is a maximum or a minimum.

You need to find the y-coordinate.

Differentiation – higher derivatives

This section deals with higher derivatives and the differentiation of trig functions.

HIGHER DERIVATIVES ○○○

When you continually differentiate a function you get higher derivatives of that function. If $f(x) = x^4 + 2x^3 - 4x^2 + 2x + 6$ find $f'(x)$, $f''(x)$ and $f'''(x)$.

> Write down $f'(x)$, $f''(x)$ and $f'''(x)$.

THE JARGON
$f'(x)$ is the first derivative of the function $f(x)$, $f''(x)$ is the second derivative, and so on. You can write dy/dx for $f'(x)$ and d^2y/dx^2 for $f''(x)$.

MAXIMUM AND MINIMUM ○○○

You can use the second derivative to find out the nature of a turning point on a curve.

> Highlight the correct term.

$$\frac{d^2y}{dx^2} < 0 \Rightarrow \quad \text{minimum, maximum, point of inflexion}$$

$$\frac{d^2y}{dx^2} > 0 \Rightarrow \quad \text{minimum, maximum, point of inflexion}$$

DIFFERENTIATING SIN X AND COS X ○○○

Differentiating $\sin x$ and $\cos x$ is easy if you remember the rules.

> Complete these equations.

$$\frac{d}{dx}(\sin x) =$$

$$\frac{d}{dx}(\cos x) =$$

If $f(x) = \sin 2x$ find $f'(x)$, $f''(x)$ and $f'''(x)$.

> The function $\sin 2x$ is a composite function. Either differentiate it directly or use the substitution $u = 2x$ and apply the chain rule.

LINKS
See the chain rule on p. 47.

Turn the page for some exam questions on this topic ➤

For more on this topic, see pages 52–53 of the *Revision Express A-level Study Guide*

EXAM QUESTION 1

If $y = 2x^3 - 3x^2 - 12x + 4$ find d^2y/dx^2, hence find the coordinates of the stationary points on the curve represented by this equation. Determine the nature of these stationary points.

DON'T FORGET
You can determine the nature of turning points by another method, but use this question to practise the second derivative.

You also need to find the y-coordinate of each stationary point.

EXAM QUESTION 2

If $y = \tan x$ show that $dy/dx = \sec^2 x$.

First write tan x in terms of sin x and cos x. Now use the quotient rule to find dy/dx. You'll need to use a trig identity to finish.

LINKS
See trigonometric equations on p. 21.

Integration – further functions

Having mastered the basics, try integrating more complicated functions and finding volumes of revolution.

SUBSTITUTION ○○○

Many functions can be integrated by using a substitution.

Find the integral of $(3x + 1)^6$ by using the substitution $u = 3x + 1$.

Step 1
Find du/dx and hence dx/du.

Step 2
Make the substitution and replace dx with $(dx/du)du$.

Step 3
Now everything should be in terms of u and du, so it can be integrated in the normal way.

Step 4
Finally substitute for u in terms of x.

> Complete step 1. Don't forget
> $$\frac{dx}{du} = \frac{1}{du/dx}$$

> Complete step 2.

> Complete step 3. Don't forget the constant of integration.

> Complete step 4. If you have a definite integral and limits are given, you need to change the limits in terms of u and leave out step 4.

VOLUME OF REVOLUTION ○○○

The volume of revolution of a curve about an axis can be found from these integrals:

> Write down the two integrals.

For rotation about the x-axis

For rotation about the y-axis

Find the volume generated when the area enclosed by $y = x^2$, $x = 0$, $x = 2$ and $y = 0$ is rotated through $360°$ about $y = 0$.

EXAMINER'S SECRETS
It's often helpful to leave π in your answer.

Turn the page for some exam questions on this topic ➤

For more on this topic, see pages 58–61 of the *Revision Express A-level Study Guide*

EXAM QUESTION 1

Using the substitution $u = x^3 + 2$, show how $\int_0^1 x^2(x^3 + 2)^2 dx$ can be rewritten as $\frac{1}{3}\int_2^3 u^2 du$. Hence find the volume generated when the area enclosed by $y = x^2(x^3 + 2)$, $x = 0$, $x = 1$ and $y = 0$ is rotated through $360°$ about $y = 0$.

EXAM QUESTION 2

Find the area bounded by $y = 1\sqrt{2x + 1}$, $x = 4$, $x = 12$ and $y = 0$.

Numerical methods – change of sign

When you cannot solve an equation exactly you have to rely on numerical methods to find an approximate solution.

CHANGE OF SIGN ○○○

Locate where a root of a function lies by looking for a sign change.
Show that the function $f(x) = x^3 - 3x^2 + 1$ has a root between $x = 0$ and $x = 1$ when $f(x) = 0$.

Look for a sign change.

THE JARGON
A root of a function $f(x)$ is a value of x such that $f(x) = 0$.

When looking for a root in the interval $x = a$ to $x = b$, all you need to do is show that $f(a)$ has a different sign to $f(b)$.

DECIMAL SEARCH ○○○

Once you've located where a root lies, you can find an approximation for the root by using a decimal search.

For the function $f(x) = x^3 - 3x^2 + 1$, find to 1 d.p. the root of $f(x) = 0$ which lies in the interval $x = 0$ and $x = 1$.

We have $f(0) = 1$ and $f(1) = -1$ so choose $x_1 = 0.5$

Complete the stages then highlight 'too big' or 'too small', whichever is correct. At each iteration increase x by 0.1 until you find the solution.

$f(x_1)$	=	too big/too small
x_2	=	
$f(x_2)$	=	too big/too small
x_3	=	
$f(x_3)$	=	too big/too small
x_4	=	
$f(x_1)$	=	too big/too small

At each stage you need to decide whether your approximation is too big or too small. If the value is > 0 and the interval changes sign from $-$ to $+$ then the value is too big. If the value is > 0 and the interval changes sign from $+$ to $-$ then the value is too small.

DON'T FORGET
$x_{n+1} = x_n - f(x_n)/f'(x_n)$

NEWTON–RAPHSON ○○○

Newton–Raphson is far more efficient than interval bisection.

Given that $f(x) = x^3 - x^2 + 20x - 5$ has a root between $x = 0$ and $x = 1$, perform one iteration of the Newton–Raphson method to find the root to 1 d.p. using 0.5 as your first approximation.

Complete the working.

Begin with $f'(x) = 3x^2 - 2x + 20$

x_1	=
$f(0.5)$	=
$f'(0.5)$	=
x_2	=

LINKS
See differentiation on p. 29.

Turn the page for some exam questions on this topic ➤

EXAM QUESTION

● ● ●

The sketch graph shows $f(x) = x^3 - 8x^2 - 4x - 5$. It crosses the x-axis at s. The consecutive integers a and b are either side of s.

Find the values of a and b. Use the Newton–Raphson method to find the value of s to four significant figures.

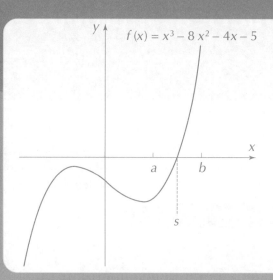

First find a and b by looking for a sign change

When answering this question you will need to look for a sign change. From the graph can you guess a possible value for s? Once you have found a and b, the midpoint is a sensible first approximation for Newton–Raphson.

Now use the Newton–Raphson method

WATCH OUT
Work to a sensible level of accuracy. The question asks you for four significant figures so work to at least five.

Numerical methods – convergence

There are other ways of finding numerical solutions to equations. Not all numerical methods work. Sometimes a method will not give a solution but will lead to divergence.

REARRANGING IN THE FORM $x = g(x)$ ○○○

This method often finds a root of an equation $f(x) = 0$ by rewriting it as $x = g(x)$.

The root α of the equation $x^3 - 4x - 7 = 0$ can be found by using the iterative formula

> Try out this formula using your calculator.

$$x_{n+1} = \sqrt[3]{4x_n + 7}$$

Taking $x_1 = 2.5$ find x_8 and write down α correct to 4 s.f.

> To show that an iterative formula finds the root of a given equation, write it without the subscripts n and $n + 1$. Now rearrange until you get the equation you're asked to solve.

STAIRCASE AND COBWEB DIAGRAMS   ○○○

Not all rearrangements work, so be able to recognize convergence and divergence. They can be represented graphically.

> Below each diagram say whether it shows divergence or convergence to the root α.

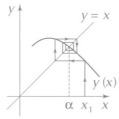

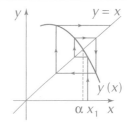

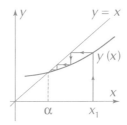

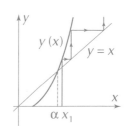

Turn the page for some exam questions on this topic ➤

EXAM QUESTION 1

● ● ●

Show that the iterative formula

$$x_{n+1} = \sqrt[3]{\frac{15 - x_n}{x_n^2}}$$

can find a root of the equation $x^5 + x - 15 = 0$ and that the equation has a root between $x = 1$ and $x = 2$. Start with $x_1 = 2$ then use the formula to find $x_2, x_3, \ldots, x_6$ giving your answers to five decimal places. Say whether this sequence is divergent or convergent.

You will need to rearrange the iterative formula until you get the original equation. Look for a sign change to see that the interval contains a root.

LINKS
See change of sign on p. 53.

Now you can investigate to see whether the sequence is diverging or converging.

EXAM QUESTION 2

● ● ●

Use rearrangement to decide which of these iterative formulae could be tried when finding a root of the equation $x^4 - 2x^2 + x - 1 = 0$.

(a) $x_{n+1} = \sqrt{\dfrac{1 - x_n}{x_n^2 - 2}}$　　(b) $x_{n+1} = \sqrt{\dfrac{x_n}{x_n^2 + 2}}$　　(c) $x_{n+1} = \dfrac{1}{x_n^3 - 2x_n + 1}$

Vectors

Vectors have magnitude and direction; scalars, have only magnitude. For example, a speed of $5\,\mathrm{m\,s^{-1}}$ is a scalar (it only has magnitude) but a velocity of $5\,\mathrm{m\,s^{-1}}$ vertically downwards is a vector (it has magnitude and direction).

REPRESENTING VECTORS   ○○○

You can use a line with an arrow to represent a vector. The length of the line represents the magnitude and the arrow shows the direction of the vector.

> Draw lines to represent the following vectors: (a) a velocity v magnitude 3 $\mathrm{m\,s^{-1}}$ and direction north-east, (b) a displacement AB of magnitude 4 m and direction 30° below the horizontal.

RESULTANT VECTORS   ○○○

A resultant vector is the sum of two or more vectors. You can find the resultant by using a vector triangle.

Two forces of 4 N and 2 N have an angle of 40° between them. Find to 2 d.p. the magnitude of the resultant vector.

> The angle between two vectors is the angle made when both vectors are drawn facing away from each other and starting from the same point. To make a vector triangle, redraw the vectors 'nose to tail' so that one vector starts where the other vector finishes. The resultant vector is the third side of the triangle, and the angle between the vectors is the supplementary angle. Find the magnitude of the resultant vector by using the cosine rule.

Find to 2 d.p the magnitude and direction of the vector $6\mathbf{i} - 2\mathbf{j}$.

> If the vector is described by components i and j, use Pythagoras for its magnitude and trigonometry for its direction angle.

RESOLVING VECTORS   ○○○

You need to be able to resolve a vector into two perpendicular components. F is a force of 4 N at 30° to the horizontal. Resolve F into horizontal and vertical components.

> To resolve a vector, draw a right-angled triangle with the vector as the hypotenuse. Line up the other sides of the triangle with the directions in which you're resolving; use trigonometry to find the component sides of the triangle.

Turn the page for some exam questions on this topic ➤

EXAM QUESTION 1 ● ● ●

The diagram shows the forces acting on a body. Resolve the forces parallel to *OX* and then parallel to *OY*. Hence find the resultant force on the body and state its direction, giving your answer to 2 d.p.

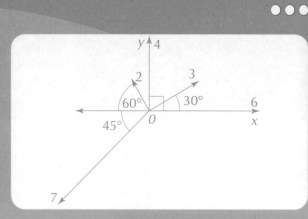

Find the total force in the *x*-direction and the total force in the *y*-direction.

Now find the resultant using Pythagoras and the direction using trigonometry. A sketch will help at this stage.

EXAM QUESTION 2 ● ● ●

Two boys are pulling a sled using two ropes attached to the same point on the sled. The ropes are parallel with the ground and make an angle of 50° with each other. If the tensions in the rope are 5 N and 3 N, what is the magnitude of the resultant pulling force?

Draw a sketch and a vector triangle.

Kinematics

Kinematics is the study of motion. You are considering a body moving in a straight line with a constant acceleration a. The body starts with an initial velocity u and finishes with a final velocity v, covering a displacement s in a time t.

GRAPHS

The motion of a body can be described by a displacement–time graph or a velocity–time graph.

> Connect each type of graph on the left with its corresponding statements on the right.

1 Displacement–time graph   Gradient = velocity

2 Velocity–time graph   Gradient = acceleration

                                    Area = displacement

Sarah throws a ball vertically up into the air and catches it again. Which of these graphs could be the velocity–time graph for the ball?

> Circle the correct letter.

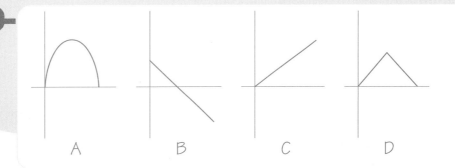

A   B   C   D

DON'T FORGET
Acceleration is constant here.

EQUATIONS OF MOTION

The four equations of motion relate the variables u, v, s, t and a.

> Write down the four equations of motion.

DON'T FORGET
Each of the equations leaves out one of the variables.

VERTICAL MOTION UNDER GRAVITY

You are applying the equations of motion vertically in a straight line. The only acceleration is the constant acceleration due to gravity g.

> Write down the equation of motion needed to find the time taken for a ball to reach its greatest height. What can you say about the value of v at this height?

Turn the page for some exam questions on this topic ➤

EXAM QUESTION 1

● ● ●

DON'T FORGET
Distance is not the same as displacement. Displacement is a vector, so it can be positive or negative; distance is a scalar.

A particle is moving with a velocity of $8\,\mathrm{m\,s^{-1}}$ in a straight line. At a time $t = 0$ it is subjected to an acceleration of $-4\,\mathrm{m\,s^{-2}}$ for 4 s and then continues at a constant velocity for 3 s before being brought to rest by a constant deceleration in 6 s. Draw a velocity–time graph then find (a) the total distance covered and (b) the total increase in displacement.

Your graph should use straight lines only. Displacement on a velocity–time graph is the area under the curve.

(a) Find the total distance covered

(b) Find the increase in displacement

EXAM QUESTION 2

● ● ●

A particle is projected upwards with a speed of $16\,\mathrm{m\,s^{-1}}$ from point A. At the same time a second particle is released from rest 8 m directly above A. Find d, the distance above A, and the time taken for the particles to collide (take $g = 9.8\,\mathrm{m\,s^{-2}}$).

If d is the distance above A where the particles collide, then for the second particle $s = 8 - d$.

Now solve the simultaneous equations.

60

MATHS

Force

Forces are very important in mechanics. A force is necessary to make an object begin to move or to bring a moving object to rest. Forces have magnitude and direction and are therefore vectors. The unit of force is the newton (N).

DIFFERENT TYPES OF FORCE   ○○○

Name these common forces.

Force due to gravity

Forces found in strings and springs

Contact forces

FORCE DIAGRAMS   ○○○

Many mechanics questions can be answered using a force diagram.

A block of weight W rests on a smooth slope inclined at 20° to the horizontal. The block is held at rest by a rope parallel to the slope. Which diagram correctly shows all the forces acting on the block?

Circle the correct letter.

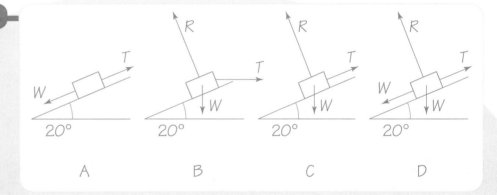

A   B   C   D

Use filled arrows to represent the forces and label them with appropriate letters, e.g. W for weight and F for friction.

Draw a force diagram to represent a book resting on a table. The book is subjected to a horizontal force P and the book is at rest.

Draw a force diagram to represent a person standing in a lift that is accelerating upwards.

Turn the page for some exam questions on this topic ➤

EXAM QUESTION 1

● ● ●

The diagram shows a rough plank resting on a block with one end of the plank on rough ground.

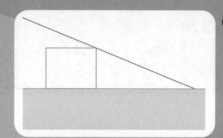

Draw diagrams to show

(a) the forces acting on the plank

(b) the forces acting on the block

These questions would only be the start of an exam question. You would probably be asked to find the magnitude of some of the forces when given some more information.

(a) Forces acting on the plank

(b) Forces acting on the block

EXAM QUESTION 2

● ● ●

Two bricks A and B have weights 15 N and 7 N respectively. B rests on a table and A rests on B. Draw diagrams to show the forces acting (a) on brick A and (b) on brick B.

(a) Forces on brick A

(b) Forces on brick B

Newton's laws of motion

Isaac Newton's laws form the basis of mechanics. You need to be able to quote these laws and use them in solving problems.

NEWTON'S THREE LAWS   ○○○

Newton's first law

> Write down Newton's three laws in the spaces provided.

Newton's second law

Newton's third law

$F = ma$ ○○○

A car of mass 1400 kg is pushed by a force of 200 N. Calculate the acceleration of the car.

> Most of your calculations will use Newton's second law. Use Newton's second law and rearrange to find a.

DON'T FORGET
Weight = mg

DON'T FORGET
You can use a double-arrowhead to represent acceleration.

Here is a force diagram of a person of mass M kg standing in a lift. The lift is moving vertically upwards with a constant acceleration.

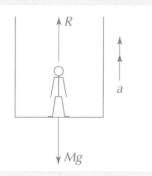

Which equation correctly describes the motion?

(A) $R = Mg$

(B) $Mg - R = Ma$

(C) $R - Mg = Ma$

(D) $Ma - R = Mg$

Turn the page for some exam questions on this topic ➤

EXAM QUESTION 1

● ● ●

A car of mass 450 kg is brought to rest in a time of 5 s from a speed of 25 m s^{-1}.

Assuming the braking force is constant and assuming there is no resistance to motion, find the force exerted by the brakes.

You'll need to find *a* by using a constant acceleration equation. A diagram will help you decide the directions for the braking force and the acceleration.

THE JARGON
The assumptions made in a model are the simplifications such as no air resistance or lack of friction.

LINKS

See kinematics on p. 59.

EXAMINER'S SECRETS
When dealing with connected particles, always consider each particle separately. A carefully labelled diagram is essential for answering this question.

EXAM QUESTION 2

● ● ● ●

Two particles of mass 3 kg and 7 kg are connected by a light inelastic string passing over a smooth fixed pulley. The system is released from rest.

Say how these two assumptions affect the model: (a) the pulley is smooth and (b) the string is inelastic.

Now find the acceleration of the particles and the tension in the string; leave *g* in your answer.

First draw a diagram.

Using *F* = *ma* write down two equations, one for each particle, then solve them simultaneously.

Projectiles

Any body moving only under the action of its own weight after being given an initial velocity u is a projectile. We consider bodies projected at some given angle to the horizontal.

EQUATIONS ◯◯◯

LINKS
See kinematics on p. 59.

Resolve the motion into the x-direction (horizontal) and the y-direction (vertical). In the x-direction things are quite simple as there is no acceleration. In the y-direction there is acceleration due to gravity.

> What is the component of the initial velocity u in the x-direction? Hence write down an equation for x, the displacement in the x-direction.

Equations in the x-direction

> What is the component of the initial velocity u in the y-direction? Hence, using the equations of motion, write down two equations in the y-direction, one for the displacement y and one for the velocity v_y after time t.

Equations in the y-direction

MAXIMUM HEIGHT AND RANGE ◯◯◯

Many problems involve finding the greatest height and the range (horizontal distance) of the projectile.

A particle is projected with a velocity of $40\,\mathrm{m\,s^{-1}}$ at an angle of $30°$ to the horizontal. What is the greatest height reached by the particle and what is the range? Take $g = 10\,\mathrm{m\,s^{-2}}$.

> For the greatest height you only need to think in the y-direction and let $v = 0$.

Greatest height

> Find t and substitute into the displacement equation.

> For the range, the time in the air will be twice the time taken to reach the greatest height, something you've already found. Substitute this time into the equation for x.

Range

DON'T FORGET
The maximum range of a projectile is found when $\theta = 45°$.

Turn the page for some exam questions on this topic ➤

EXAM QUESTION 1

A particle is projected from a point O, with initial speed V m s^{-1} at an angle θ above the horizontal. After 8 s the vertical component of the velocity is 10 m s^{-1} downwards and the horizontal component is 18 m s^{-1}. Find to 2 d.p. the values of V and θ (take $g = 9.8$ m s^{-2}).

> Find the initial velocity in the y-direction by using $v = u + at$.

> You already know the initial velocity in the x-direction because it stays constant.

> Find V by using Pythagoras.

> You can find θ by using trigonometry.

EXAM QUESTION 2

A particle is projected with speed 20 m s^{-1} at 30° to the horizontal. Find to 1 d.p. the velocity of the particle after 4 s ($g = 9.8$ m s^{-2})

DON'T FORGET

The final velocity will be the resultant of the vertical and horizontal components. When you have found these components you must combine them using Pythagoras to obtain the final velocity. You must also state the direction in which the particle is travelling (use trigonometry).

Friction

Friction is the 'sticky' force that occurs whenever one surface slides over another. In mechanics, when modelling with friction, we say that the contact is rough.

THE FRICTION FORMULA   ○○○

Say how friction acts in relation to the way an object is trying to move.

Write an inequality for friction relating the friction F, normal reaction N and the coefficient of friction μ.

Explain why this is an inequality.

LIMITING FRICTION   ○○○

Only when friction is limiting will an object start to move.

A book of weight 4N rests on a rough table. It is pushed by a horizontal force P and the coefficient of friction is 0.75. Find the value of the frictional force and describe the motion for $P = 1$, $P = 3$ and $P = 5$.

EXAMINER'S SECRETS
Always start by drawing a good force diagram. You will need to find the normal contact force.

LINKS
See force on p. 61.

Turn the page for some exam questions on this topic ➤

EXAM QUESTION 1

● ● ●

A toboggan of weight 80 N rests on a snow-covered slope inclined at an angle of 15° to the horizontal. Given that the coefficient of friction between the toboggan and the slope is 0.2, find whether the toboggan will slide. Give your answers to 2 d.p.

Draw a force diagram. Decide which force will make the toboggan slide and determine whether or not this force is greater than the limiting friction.

EXAM QUESTION 2

● ● ●

A suitcase of mass 40 kg is pulled using a rope inclined at 30° to the horizontal. If the coefficient of friction between the suitcase and the floor is 0.25, what is the least force needed to make the suitcase move? Take $g = 9.81$ m s^{-2} and give your answers to 2 d.p.

EXAMINER'S SECRETS
Always determine whether or not the body is moving. If it is not moving, you must decide whether the friction is limiting.

Moments and equilibrium

When forces are in equilibrium the resultant is zero and there is no turning effect.

LAMI'S THEOREM ○○○

Finish this sentence.

Lami's theorem can only be used when

A weight of 20 N is hanging from two strings inclined at 30° and 60° to the vertical. What are the tensions in the strings? Answer to 2 d.p.

Draw a good force diagram and work out any missing angles. Write Lami's theorem then rearrange to find the tensions.

Using the sine rule

If there is more than one tension in a question, the chances are that they have different magnitudes. Give them different labels, e.g. T_1 and T_2.

EQUILIBRIUM PROBLEMS ○○○

If more than three forces are present then resolving the forces and taking the resultant force as zero can solve equilibrium problems.

A particle of weight 5 N rests on a rough slope inclined at 30° to the horizontal. The coefficient of friction is 0.23 and the particle is prevented from sliding by an additional force H parallel to the slope. Given that the particle is on the point of sliding, find the value of H to 2 d.p.

First draw a force diagram then resolve parallel and perpendicular to the slope.

LINKS
See friction p. 67.

Resolving forces parallel to the slope

Resolving forces perpendicular to the slope

MOMENTS ○○○

Complete this definition.

The moment of a force about an axis is

Turn the page for some exam questions on this topic ➤

For more on this topic, see pages 174–175 and 180–181 of the *Revision Express A-level Study Guide*

EXAM QUESTION 1

A uniform rod *AB* of length 2 m and weight 30 N rests horizontally on smooth supports at *A* and *B*. A weight of 5 N is attached to the rod at a distance of 0.3 m from *A*. Find to 2 d.p. the forces exerted on the rod by the supports.

Taking moments about appropriate axes can easily solve this problem. Choose an axis that eliminates one of the unknowns from your equation.

DON'T FORGET
The normal reactions at the two ends of the rod are not the same.

EXAM QUESTION 2

A ladder, which is 4 m long and has a mass of 20 kg, leans against a smooth vertical wall with its foot in contact with a rough horizontal floor. The ladder makes an angle of 45° with the horizontal and is on the point of slipping. Find the reaction between the wall and the ladder and find the coefficient of friction between the ladder and the floor. Take $g = 9.8 \, \text{m s}^{-2}$.

Draw a good force diagram showing dimensions as well as forces.

Resolve forces into two perpendicular directions.

Remember that friction is limiting.

Take moments about a convenient axis. As the forces are in equilibrium the total moment is zero.

You should now be able to solve your equations.

Momentum and impulse

When two balls collide or a bat hits a ball, what happens to the velocities and the forces during and after the impact?

DEFINITIONS   ○○○

> Give one definition of momentum and two definitions of impulse.

DON'T FORGET
The units of momentum and impulse are newton-seconds (N s).

Momentum

Impulse: definition 1

Impulse: definition 2

IMPULSE   ○○○

DON'T FORGET
All masses must be in kilograms. Convert 80 g to 0.08 kg.

DON'T FORGET
There are two forms of the impulse equation. You may need to use one or other, or both.

With questions involving impulse you may need to carry out calculations with velocities or with force.

A bat strikes a cricket ball of mass 80 grams giving it a velocity of $45 \, \text{m s}^{-1}$. Find the impulse on the ball.

A particle of mass 3 kg is travelling with a velocity of $6 \, \text{m s}^{-1}$. What constant force acting in the same direction of the particle will increase its speed to $30 \, \text{m s}^{-1}$ in 4 s?

CONSERVATION OF MOMENTUM   ○○○

> Explain conservation of momentum.

A ball of mass 8 kg is moving with a velocity of $5 \, \text{m s}^{-1}$. It collides with a ball of mass 3 kg which is at rest. After the collision the 8 kg ball is moving at $2 \, \text{m s}^{-1}$ in the same direction as before. Find the velocity of the smaller ball after the collision.

> Draw a diagram showing the masses and velocities before and after the collision.

Turn the page for some exam questions on this topic ➤

EXAM QUESTION 1

● ● ●

A bullet is fired into a box of sand mounted on a trolley. The bullet has a mass of 15 grams and the mass of the trolley and sand is 4 kg. After the bullet is fired, the box with the bullet embedded in it moves off with a speed of 1.6 m s^{-1}. What was the speed of the bullet just before it hit the sand?

> The mass of the bullet and the mass of the sand need to be added together when calculating the momentum after impact.

EXAM QUESTION 2

● ● ●

THE JARGON
Coalesce means the particles stick together and become one body.

Two particles of the same mass are travelling towards each other along the same line with constant speeds 6 m s^{-1} and 2 m s^{-1}. If they collide and coalesce, find their joint speed after impact.

> If two bodies are approaching each other, you need to take one velocity as positive and the other as negative.

EXAM QUESTION 3

● ● ●

Particles A, B and C lie in a straight line and have masses 4m, 3m and m respectively. Initially B and C are at rest with A projected towards B with a speed of 5 m s^{-1}. After A collides with B, the speed of A is 2 m s^{-1} in the same direction as before. When B collides with C the particles coalesce and move off with a joint speed v. Assuming no resistance to motion, find the value of v and the speed of B just before colliding with C.

> Tackle the collisions one at a time using the conservation of momentum for each.

Graph theory

Graphs in decision and discrete mathematics are not like the (x, y) grids normally associated with graphs. They are networks with vertices (or nodes) and edges (or arcs).

TYPICAL GRAPHS USED IN GRAPH THEORY ○○○

Describe each of these graphs in the spaces provided.

Planar graphs

Simple graph

Complete graphs

WATCH OUT
There are more graphs than are noted here. You may also be expected to show the information from a network in a table.

Bipartite graphs

Weighted graphs

Isomorphic graphs

LINKS
See spanning trees on p. 83.

Trees

EULER AND HIS TRAILS ○○○

Highlight Euler's formula

Euler's formula is

$$v + e + f = 2 \qquad v - e + f = 2 \qquad f + e = v \qquad v - e - f = 1$$

REVISION EXPRESS
Finding out whether a graph is traversable by using degrees is on p. 87 of the *Revision Express A-level Study Guide*.

Traversability is where every edge of a network may be traced only once without taking pen from paper. There are two types of trail.

State whether each sentence is true or false.

An eulerian trail will finish where it starts (it is closed).

A semi-eulerian trail will not finish where it started (it is not closed).

Eulerian and semi-eulerian trails are not traversable.

K_N – A COMPLETE GRAPH WITH N VERTICES ○○○

Next to the diagram of K_5, show that K_4 is planar.

IF YOU HAVE TIME
Using Euler's formula, show that K_5 is non-planar.

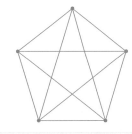

Turn the page for some exam questions on this topic ➤

EXAM QUESTION 1

● ● ●

(a) Draw the graph of K_6. How many edges does it have?

(b) How many edges does the graph of K_n have?

First draw the graph

Now work out the numbers of edges

Try to work out the relationship between the number of vertices and how many they are connected to. Also, think of how many vertices are needed to make an edge.

EXAM QUESTION 2

● ● ●

For each graph show whether it is (a) traversable with an eulerian trail, (b) traversable with a semi-eulerian trail or (c) not traversable.

By sketching over the graphs see if they are traversable. State whether they are eulerian trails, semi-eulerian trails or not traversable.

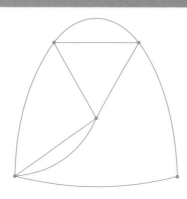
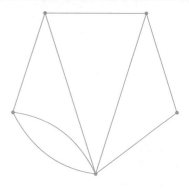
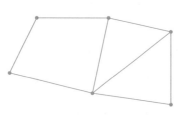
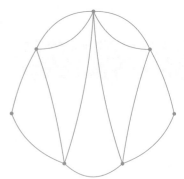

DON'T FORGET

You can use the orders of the vertices to state whether or not a graph is traversable. Try a few and see what happens. You are looking to see how many edges meet at each vertex and whether the number is odd or even.

Spanning trees

Spanning trees are connected graphs which have no circuits. Problems associated with spanning trees may be about cabling towns, for example, and finding the minimum set-up costs. To solve these problems we need to find minimum spanning trees.

ALGORITHMS FOR FINDING MINIMUM SPANNING TREES   ○○○

Prim's algorithm

State whether each sentence is true or false.

Prim's algorithm is a greedy algorithm
Prim's algorithm starts with the shortest edge
Prim's algorithm chooses any vertex to be the starting point

Kruskal's algorithm

State whether each sentence is true or false.

Kruskal's algorithm is a greedy algorithm
Kruskal's algorithm starts with the shortest edge
Kruskal's algorithm chooses any vertex to be the starting point

Prim's algorithm

Set out the steps in Prim's algorithm.

Kruskal's algorithm

Set out the steps in Kruskal's algorithm.

EXAMINER'S SECRETS
Learn both algorithms off by heart.

NETWORKS IN TABLES   ○○○

Computers can be used to find a minimum spanning tree, although the graph needs to be converted to a table first. Complete the table for this network.

	A	B	C	D
A	--	7	8	5
B				
C				
D				

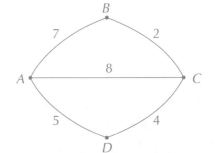

Using these steps for Prim's algorithm, delete columns in the table with a highlighter until you're left with just one column. This process will produce the edges for a minimum spanning tree.

Suppose you start at A. The shortest edge is AD, so delete columns A and D (to prevent circuits from being formed). The shortest length from A or D is DC. Therefore delete column C. The last edge would therefore be from A, D or C and is CB, length 2.

Draw the minimum spanning tree.

Minimum spanning tree

Turn the page for some exam questions on this topic ➤

EXAM QUESTION 1

A model railway is set up with places for the train to visit as shown in the network. Using (a) Prim's algorithm and (b) Kruskal's algorithm work out how to connect all the places so that the total length of railway track is a minimum, starting from A.

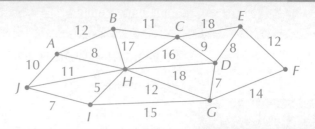

(a) Prim's algorithm

Work out the minimum spanning tree using Prim's algorithm, and write down its total length.

EXAMINER'S SECRETS
Include enough information to show the examiner you've used Prim's algorithm. The final tree on its own isn't enough.

(b) Kruskal's algorithm

Work out the minimum spanning tree using Kruskal's algorithm, and write down its total length.

DON'T FORGET
A problem may have more than one minimum spanning tree, but the minimum length will always be the same.

EXAM QUESTION 2

This table shows a complete graph K_5. How can you tell that it's a complete graph?

	P	Q	R	S	T
P	–	8	12	13	7
Q	8	–	5	10	14
R	12	5	–	6	11
S	13	10	6	–	4
T	7	14	11	4	–

Copy out the table and, using Prim's algorithm, obtain a minimum spanning tree. Draw the minimum spanning tree and write down its total length.

Draw a minimum spanning tree of the information.

Shortest paths

Shortest path problems do not require you to go along every edge nor do they expect you to visit every node. All they require is that you get from one specified node to another by the shortest route.

DIJKSTRA'S ALGORITHM TO FIND SHORTEST PATHS ○○○

Fill in the missing words to complete Dijkstra's algorithm.

Step 1 Give the starting ____ a value of ____ and then box the ____ at this ____ .

Step 2 All the vertices that are joined to the ____ boxed vertex need to be ____ labelled with the time from the ____ vertex.

Step 3 From the ____ labels that are written on the graph, choose the one with the ____ value. Make this label ____ by putting a box around it.

Step 4 Keep doing steps ____ and ____ until the vertex you are aiming to get to has been ____ . If at any later point a ____ label is reduced, use the reduced value. If it is not reduced then the label does not change.

EXAMINER'S SECRETS
When you're using the algorithm make sure you write down enough information so it's very clear that you *are* using Dijkstra's algorithm.

AN EXAMPLE USING DIJKSTRA'S ALGORITHM ○○○

Following Dijkstra's algorithm write on this network to show how to label the vertices for finding the shortest path from *A* to *G*.

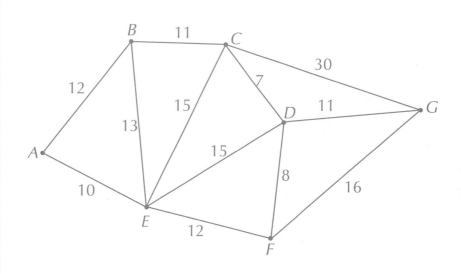

Write down the shortest path and its length.

Turn the page for some exam questions on this topic ➤

EXAM QUESTION

● ● ●

Here is a network of several towns and cities in France. The values are the distances between these places in kilometres. Work out the shortest path from Cherbourg to Marseille using Dijkstra's algorithm.

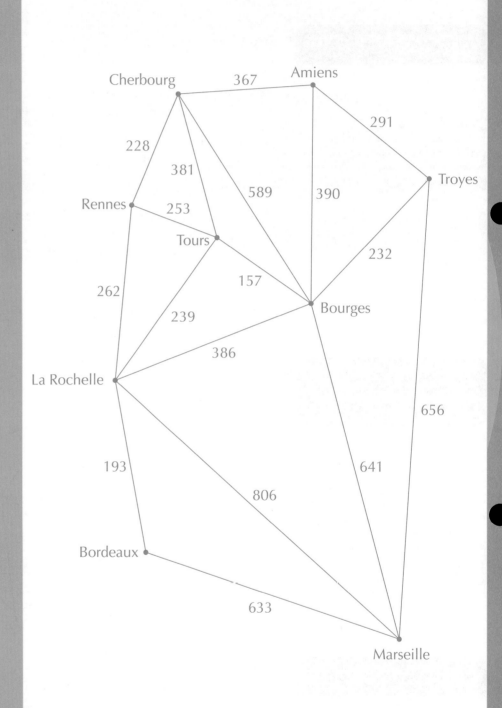

Write down the route for the shortest path and its length.

Route inspection problem

A route inspection problem involves going along every edge and returning to the original vertex. The problem can easily be understood if you think of a postman, starting at the post office, delivering letters to all streets and then returning to the post office. Mei-Ko Kwan, a Chinese mathematician, was the first person to analyse this problem. That's why it's more commonly known as the Chinese postman problem.

CHINESE POSTMAN PROBLEM ○ ○ ○

Highlight the correct answers.

The route can be accomplished without repeating a street
if all the vertices are   even/odd/mainly even.
New edges (or streets)   can be added/cannot be added.
Vertices are made even by   adding a new edge/repeating an edge.

THE CHINESE POSTMAN ALGORITHM ○ ○ ○

Fill in the spaces to complete the steps of the Chinese postman algorithm.

Step 1 Ascertain which   , if any, are   .

Step 2 Pair these   so that any extra distances added are kept to a   .

Step 3 Put in these repeated   so that the graph formed has all its   .

Step 4 Find the   trail by inspection.

Apply the algorithm to this network; start and finish at *A*.

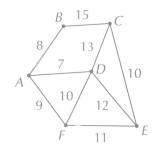

Write down the route and its length.

WATCH OUT
The route worked out is just *one* of the possible routes.

OTHER APPLICATIONS ○ ○ ○

Route inspection algorithms have many other applications.

Write down as many different applications as you can.

Turn the page for some exam questions on this topic ➤

EXAM QUESTION

Here is a network of streets along which a newspaper delivery person has to deliver newspapers. They pick up the newspapers from *A* and they need to return there with any spares at the end of their delivery. Which route should be taken to keep it to a minimum? How long is this route? Distances are in metres.

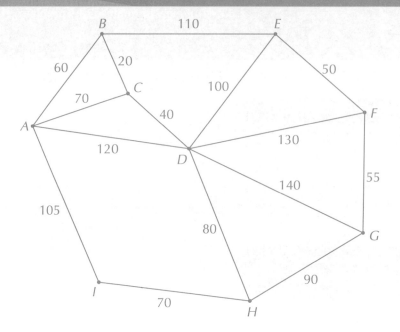

Draw out the network and add any extra edges required to complete the Chinese postman algorithm. Show the route.

DON'T FORGET
Your route is just one possible route. You could try to find some other routes.

Write out the route.

Work out the length of the route.

Travelling salesperson problem

The travelling salesperson problem (TSP) is similar to the Chinese postman problem except it involves visiting all the vertices rather than traversing all the edges. If the vertices are towns or cities and the edges are roads, rail tracks or air routes then it should be clear why it's known as TSP, with the traveller wanting to take the shortest route yet still visiting every vertex.

THE NEAREST-NEIGHBOUR ALGORITHM ○○○

Unfortunately, there is no simple algorithm that will automatically give you the shortest path. The nearest-neighbour algorithm will produce a route that is reasonable, but not necessarily the shortest.

> **Explain the nearest-neighbour algorithm.**

DON'T FORGET
There is more than one method for finding upper and lower bounds.

UPPER AND LOWER BOUNDS ○○○

The only way to be completely confident you have the optimal route for the TSP is to look at every route. Explain the reason for working out the upper and lower bounds.

> **Give a brief explanation why upper and lower bounds are calculated for the TSP.**

Let us obtain the upper bound and then the lower bound for this network.

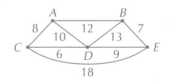

> **Draw the minimum spanning tree.**

Calculating the upper bound
Here is the minimum spanning tree for the network.

> **Redraw your minimum spanning tree with edges going both ways and write down the value of this upper bound.**

Traversing edges both ways will give an upper bound.

> **Draw the minimum spanning tree of the remaining vertices.**

Calculating the lower bound
After deleting vertex *E* (chosen randomly), we obtain the minimum spanning tree for the remaining vertices.

> **Write down the values from the minimum spanning tree and add the two shortest edges from *E*.**

Adding the two shortest edges from *E* gives a lower bound.

Turn the page for some exam questions on this topic ➤

EXAM QUESTION 1

● ● ●

Fred is trying to sell his new lawnmower invention to businesses in different towns. Here is a network of the towns Fred wants to visit and the roads connecting them. Distances are in kilometres. Fred lives at *A*. Using the nearest-neighbour algorithm find a route that starts and finishes at *A* and visits every town; write down its length.

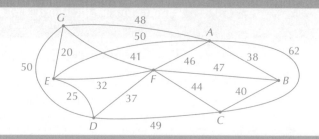

Show how to use the nearest-neighbour algorithm to work out a route.

EXAMINER'S SECRETS
Give enough detail to show the examiner you understand the algorithm.

Write down the total length of the route.

EXAM QUESTION 2

● ● ●

DON'T FORGET
There is more than one way of calculating upper and lower bounds. Be clear how you have found them and why they are an upper or lower bound.

Find an upper bound and a lower bound for the length of the optimal salesperson's tour on this network.

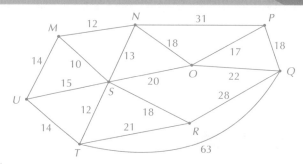

Upper bound

Lower bound (choose to omit *Q*)

Critical path analysis

Critical path analysis (CPA) helps plan and schedule projects which have many tasks; it minimizes the amount of waiting for earlier tasks to finish before later tasks can begin.

SYLLABUS CHECK
Different labelling methods are used by different boards and teachers. Check which method you are meant to use.

> Here is a table of tasks for making a cup of tea. Complete the second column with the relevant letters. If you think task B has to precede task C, then write B next to C in the 'Preceding tasks' column.

DON'T FORGET
When completing the times on a network, keep reminding yourself that it is the *earliest* possible start time and the *latest* possible finish time. Fill in the numbers working forwards and then work backwards adjusting the finish times.

> Add (1) (2) (3) (4) to the correct part of the box, where
> (1) = task or activity
> (2) = earliest possible start time
> (3) = duration of task
> (4) = latest possible finish time

> Complete the network by putting in all the required times, in minutes.

WATCH OUT
The earliest possible start time plus the duration of the activity will not always give you the latest possible finish time. You then have some float time.

> Write down a possible critical path and its duration.

NETWORKS   ○○○

A project can be split into tasks. Some of the earlier tasks must be completed before a later task can be started.

	Task	Preceding tasks
A	Water in kettle	
B	Kettle on	
C	Cups out	
D	Milk in cups	
E	Tea leaves into teapot	
F	Boiling water into teapot	
G	Pour tea into cups	

EARLIEST POSSIBLE START AND LATEST POSSIBLE FINISH   ○○○

Once the list of tasks is organized, times need to be allocated and a network set up. On a network each task is written into a box like this.

Here is a network for erecting and setting up a trailer tent.

From this network we can establish the critical path and the shortest time for completing the project.

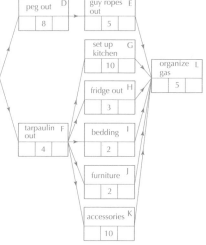

RESOURCE LEVELLING AND GANTT CHARTS   ○○○

A Gantt chart uses thick horizontal lines to represent the project tasks; it plots the days along the top of the chart. The number of people assigned to a task is often written above its line on the Gantt chart. Resource levelling is a way to make projects more cost-effective by graphing number of people against days. The graph is a vertical bar graph and resource levelling tries to reduce the variation between the heights of the bars.

Turn the page for some exam questions on this topic ➤

EXAM QUESTION

Here are the tasks from a recipe for making raspberry yogurt ice, along with their duration and the preceding tasks.

		Preceding tasks	Duration (mins)
A	Get out all the ingredients	–	10
B	Sieve raspberries, removing pips	A	5
C	Add sugar, syrup, yogurt to raspberries	B	4
D	Whisk the cream in a bowl	A	6
E	Add whisked cream to raspberry mixture	C, D	3
F	Whisk egg whites and sugar until stiff and white	A	10
G	Mix egg white into raspberry mixture	E, F	5
H	Pour into a freezer container and cover	G	7
I	Freeze	H	240

(a) Draw a network of this information; include all relevant times.

(b) Write down the critical path and the time it takes.

(c) Which task has the greatest float?

(d) What is the minimum number of people required to make this recipe in the quickest time?

Linear programming

Linear programming is used in many areas of business and manufacturing. Questions are often about maximizing profit.

LINEAR PROGRAMMING: THREE MAIN COMPONENTS   ○○○

Give a brief description for each of these three terms.

Variables

EXAMINER'S SECRETS
Linear programming questions can be quite long. Remember these three main parts and it should become easier to sort out the information.

An objective

Constraints

FORMING THE EQUALITIES AND INEQUALITIES   ○○○

A factory produces two different types of furniture: chairs and tables. The chairs need 3 hours on the lathe and 1 hour on the sprayer. The tables need 1 hour on the lathe and half an hour on the sprayer. A chair makes a £20 profit whereas a table makes an £8 profit. The lathe is available for use 14 hours per day and the sprayer for 6 hours per day. Total overheads for the factory are £50 per day. How many chairs and tables need to be made per day to maximize profits?

Write down the variables.

Write down the objective.

Write down the constraints.

PLOTTING THE GRAPH   ○○○

Using a ruler and keeping it parallel to the line $20c + 8t - 50 = 0$, find the point on the graph that shows the maximum profit.

DON'T FORGET
The maximum profit will be when the ruler is furthest away from the origin and still within the feasible region.

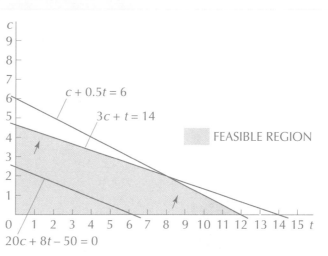

Work out the maximum profit and write down how many chairs and tables need to be made.

Turn the page for some exam questions on this topic ➤

EXAM QUESTION

● ● ●

A manufacturer produces Welsh dressers in two types, A and B. Type A requires 4 hours on the lathe, 2 hours to be assembled and 1 hour on the varnisher. It makes a profit of £120 per dresser. Type B requires 2 hours on the lathe, 1 hour to be assembled and $1\frac{1}{2}$ hours on the varnisher. Type B makes a profit of £70 per dresser. The lathe is in operation for 16 hours per day, the assemblers are available for 12 hours per day and the varnisher is available for 6 hours per day. Total overheads are £80 per day. How many of each type should be made to maximize profits?

Write down the variables, the objective and the constraints.

Now draw the graph.

Write down the number of type A dressers and the number of type B dressers that need to be made for maximum profit; work out the profit per day.

Descriptive statistics

The tasks in this section are just to consolidate your GCSE data handling. There's still lots to remember, but your calculator will help if you use it carefully.

MEAN, MODE, MEDIAN AND RANGE ○○○

Look at this frequency table then calculate the mean, mode, median and range of the scores.

Score (x)	1	2	3	4	5	6
Frequency (f)	10	18	8	8	7	9

Modal score =

Range =

Mean $= \sum fx \Big/ \sum f$

=

=

Median =

DON'T FORGET
$\sum$ means the sum of.

DON'T FORGET
When there's an even number of values, the median is the mean of the middle two.

GROUPED DATA ○○○

From this grouped frequency data, calculate an estimate of the mean.

Weight (kg)	$2.5 \leq w < 3$	$3 \leq w < 3.5$	$3.5 \leq w < 4$	$4 \leq w < 4.5$	$4.5 \leq w \leq 5$
Frequency	14	40	23	21	17

DON'T FORGET
Use class midpoints to help calculate an estimate of the mean.

STANDARD DEVIATION ○○○

Write out the formulae for calculating standard deviation.

For individual data

For a frequency distribution

Find the mean and standard deviation of 10, 11, 15, 16, 18, 20.

Mean

Variance

DON'T FORGET
The standard deviation is the square root of the variance.

Standard deviation

Turn the page for some exam questions on this topic ➤

EXAM QUESTION 1

● ● ●

The ages of the ten people working at an office are 18, 20, 23, 23, 29, 32, 33, 38, 40, 48. Find the mean and the standard deviation of their ages. What was the mean and standard deviation of their ages three years ago?

Mean and standard deviation now

Mean and standard deviation three years ago

Remember that standard deviation is a measure of spread. Was the spread of ages different three years ago?

EXAM QUESTION 2

● ● ● ●

Two squash players counted the number of shots they played in each rally. They recorded their results in a grouped frequency table.

Use the extra columns if you wish.

Shots	Rallies			
1–10	5			
11–20	19			
21–30	16			
31–40	10			
Total	50			

Find estimates for (a) the median, (b) the mean and (c) the standard deviation of the number of shots.

(a) Estimate the median

DON'T FORGET
To estimate the median, draw a cumulative frequency polygon.

(b) Estimate the mean

(c) Estimate the standard deviation

Statistical diagrams

This section looks at ways to present raw and grouped data. Just be careful with choice of grouping and class boundaries.

RAW DATA ○○○

Draw a stem and leaf diagram for this data, and from your diagram find the median.

Stem and leaf diagrams display the raw data, but give an idea of spread

17, 9, 11, 23, 12, 20, 13,

22, 14, 11, 21, 8, 19

2 | 3 means 23

CUMULATIVE FREQUENCY ○○○

Cumulative frequency looks at the spread of data. This table gives the heights of seedlings correct to the nearest centimetre.

Complete this table and draw a cumulative frequency polygon – you will need to use graph paper to make sure it's accurate.

DON'T FORGET
Plot the cumulative frequencies against the upper class boundaries (UCBs)

Height (cm)	Frequency	UCB	Cum. freq.
3–6	10		
7–10	19		
11–14	49		
15–18	59		
19–22	27		
23–26	16		

Use your cumulative frequency polygon to estimate the median and interquartile range, and draw a box plot using 2.5 as the min and 26.5 as the max.

HISTOGRAMS ○○○

If the class widths are different then use frequency density. Complete the table and draw the histogram.

Length (mm)	Frequency	Class width	Freq. density
1–20	5		
21–23	12		
24–26	28		
27–30	5		

Turn the page for some exam questions on this topic ➤

EXAM QUESTION 1 ●●●

DON'T FORGET
A cumulative frequency polygon is made by joining the points with straight lines.

This table summarizes the weights of 250 people; the weights are in kilograms to the nearest 100 g. Complete the empty columns. Represent the data on a cumulative frequency polygon. Make an estimate of the weight exceeded by 20% of the people.

Weight (kg)	No. of people	UCB	Cum. freq.
44.0–47.9	3		
48.0–51.9	17		
52.0–55.9	50		
56.0–57.9	45		
58.0–59.9	46		
60.0–63.9	57		
64.0–67.9	23		
68.0–71.9	9		

EXAM QUESTION 2 ●●●

A doctor was looking at consultation times. The doctor timed 250 appointments, measured to the nearest minute. Use the table to help construct a histogram for the data.

Time (min)	No. of appointments		
2–3	30		
4	46		
5	48		
6–7	84		
8–10	27		
11–	15		

DON'T FORGET
The boundaries for the first class are 1.5 min and 3.5 min.

In the last row you must pick a sensible value for the maximum.

Collection of data

Why do we collect data and what types can we collect?

POPULATION AND SAMPLE ○○○

Complete the definitions.

Population

Sample

Finite population

Infinite population

Countably infinite population

What are the advantages of sampling?

What are the disadvantages of sampling?

Try to list at least two advantages and two disadvantages.

TYPES OF DATA ○○○

Complete the definitions.

Qualitative data is
Quantitative data is
Discrete data takes
Continuous data can
Primary data is
Secondary data is
Height
Eye colour
Size of family
Cost in pence
Volume

Describe each type of data using one word from each of these two pairs: qualitative and quantitative, discrete and continuous. Just pick one word for eye colour

SAMPLING METHODS ○○○

Link the method on the left with its definition on the right.

1 Random

This involves taking items at regular intervals, e.g. every fifth tree when sampling in a forest.

2 Systematic

This is used to ensure the sample is representative of the population; the quota sampling method is often used instead.

3 Stratified

This is where each individual must have an equal chance of being chosen.

Turn the page for some exam questions on this topic ➤

EXAM QUESTION 1

Here is the split of year groups in a secondary school. To conduct a survey, they want a sample of 100 students. How many students from each year group are needed in order to get the correct proportional representation?

Think carefully how to get the correct number from each year group.

Year 7	Year 8	Year 9	Year 10	Year 11	Year 12
146	172	158	124	110	70

EXAM QUESTION 2

A survey is carried out to investigate the quality of new houses. All the people in the country who have recently bought a new house are to be surveyed.

(a) Give one reason why a pilot survey might be carried out first.
(b) Give one reason why the survey should be carried out by post.
(c) Give one possible disadvantage of a postal survey.

EXAMINER'S SECRETS
Be brief; don't write an essay.

EXAM QUESTION 3

(a) State a variable about a car that is discrete.
(b) State a variable about a car that is qualitative.

EXAM QUESTION 4

(a) A survey carried out inside a newsagent found that 85% of the population buy a newspaper. Why was this a poor sampling method?
(b) A phone poll carried out at 11 am on a Sunday morning revealed that less than 3% of the population regularly go to church. Why was this a poor sampling method?
(c) Some 52% of the population were estimated to watch the six o'clock news each evening after a survey was carried out at a fitness club. Why was this a poor sampling method?

Random variables

Understand and interpret notation for random variables.

DISCRETE RANDOM VARIABLES (DRVs)

○○○

Let X be the number of heads when two coins are tossed. Find the probability distribution for X and state why this is a DRV.

> Complete the calculations, then complete the table, and remember to check that the probabilities add up to 1.

$p(0 \text{ heads}) =$

$p(1 \text{ head}) =$

$p(2 \text{ heads}) =$

DON'T FORGET
The probability distribution is simply a table showing all the probabilities.

x	0	1	2
$p(X = x)$	0.25		

EXPECTATION AND VARIANCE OF A DRV

○○○

You need to know how to calculate the expected value $E(X)$ and variance $\text{Var}(X)$.

> Fill in the formulae.

$E(X) =$ $\qquad\qquad$ $\{E(X)\}^2 =$

DON'T FORGET
Always define your random variable if it hasn't been defined in the question.

$\text{Var}(X) =$

Consider a biased die. Let X be the score shown when the die is thrown. Here is the probability distribution.

x	1	2	3	4	5	6
$p(X = x)$	0.1	0.1	0.1	0.3	0.2	0.2

Find the expected score and the variance when the die is rolled.

> Complete the calculations.

$E(X) =$

SYLLABUS CHECK
The notation you are used to could well be different. Stick with what you know.

$\text{Var}(X) =$

PROBABILITY FUNCTIONS

○○○

A probability function is a simple function for calculating probabilities. A spinner can take the values 0, 1, 2, 3, 4 and the probability function is given by $p(X = x) = \frac{1}{20}(x^2 - 2x + 2)$. Find the probability distribution for X.

When $x = 0$, $p(X = 0) = \frac{1}{20}[(0)^2 - 2(0) + 2] = 0.1$

When $x = 1$, $p(X = 1) = \frac{1}{20}[(1)^2 - 2(1) + 2] = 0.05$

> Complete the table of probabilities and confirm that it's a DRV.

x	0	1	2	3	4
$p(X = x)$	0.1	0.05			

If this spinner is used at a fairground and you win a prize for spinning less than 3, the probability can be written $F(t) = p(X < t)$.

> Work out the probability of winning.

$p(X < 3) =$

Turn the page for some exam questions on this topic ➤

EXAM QUESTION 1

The discrete random variable X has the following distribution.

x	3	6	9
$p(X = x)$	$\frac{1}{2}$	$2k$	k

Find the constant k and the mean value of X.

DON'T FORGET
The probabilities add up to one.

EXAM QUESTION 2

Write a probability distribution for the number of tails when a coin is tossed four times. Find the probability of getting more than two tails.

First define your DRV then work out the probabilities, and finally draw up a table.

EXAMINER'S SECRETS
Notice the symmetry of the table when $p = \frac{1}{2}$

LINKS
See the binomial distribution on p. 95.

EXAM QUESTION 3

There are eight coins in a bag, one £1, three 50p, two 20p and two 10p. One coin is drawn at random. Write a probability distribution then calculate the expected amount and the standard deviation.

Discrete probability distributions

Ninety-seven percent of the chocolates produced by a sweet factory are suitable for sale. What is the probability that 10 chocolates will be faulty in a batch of 100?

PERMUTATIONS AND COMBINATIONS ○○○

Write the formulae for ^nP_r and ^nC_r in full.

How do you know which one to use?

Begin by deciding whether the order is important.

How many ways are there of arranging three letters from MATHS?

How many ways are there to select any 3 people from a group of 9?

BINOMIAL DISTRIBUTION ○○○

Write the formulae for the binomial distribution then have a go at the question.

For $X \sim B(n, p)$

$P(X = r) =$ $E(X) =$ $Var(X) =$

A die is rolled 5 times. What is the probability of getting 2 sixes? Why is the binomial model best here?

If you have time find P(4 or more sixes) = P(4) + P(5).

POISSON DISTRIBUTION ○○○

The Poisson distribution works with events that are randomly scattered and independent of each other. It is written $X \sim Po(\lambda)$ where λ is the mean.

Write the formulae for the Poisson distribution then have a go at the question.

$P(X = r) = \lambda^r e^{-\lambda}/r!$ $E(X) = \lambda$ $Var(X) = \lambda$

A football team scores an average of 1.2 goals per game. Find the probability that in their next game they score (a) 3 goals, (b) less than their average.

POISSON APPROXIMATION TO THE BINOMIAL ○○○

Finish the calculations to this question. Ninety-seven percent of the chocolates produced by a sweet factory are suitable for sale. What is the probability that 10 chocolates will be faulty in a batch of 100?

For a large number of trials and small probability, the Poisson distribution provides an approximation to the binomial using $\lambda = np$.

In this question $X | B(100, 0.03)$ is approximated by the Poisson using $\lambda = np = 100 \times 0.03 = 3$, therefore $X \sim Po(3)$

Turn the page for some exam questions on this topic ➤

EXAM QUESTION 1

Use the letters from the word EXAMINATION to obtain (a) the number of different arrangements, (b) the number of different arrangements that begin with X and (c) the number of different arrangements that begin and end in A.

Take care over the repeated letters.

(a) Number of different arrangements

(b) Number of different arrangements that begin with X

(c) Number of different arrangements that begin and end in A

EXAM QUESTION 2

Four fair coins are tossed and the total number of heads showing is counted. Find the probability of obtaining (a) only 1 head, (b) at least 1 head, (c) the same number of heads as tails.

Use the binomial model.

P(at least 1 head) = 1 − P(no heads)

EXAM QUESTION 3

The number of phone calls received at a switchboard on a weekday afternoon follows a Poisson distribution with a mean of 7 calls per five-minute period.

Find the probability that (a) there are no calls in the next five minutes, (b) four calls are received in the next five minutes, (c) more than two calls are received between 3:35 and 3:40.

Start by deciding on the value of λ.

For part (c) do something similar to Question 2(b).

Normal distribution

The normal distribution has many examples that you need to learn.

STANDARDIZING DATA ○○○

This begins as a simple method of comparison. Consider two test scores: 64 in maths (mean 54 and standard deviation 5) and 78 in biology (mean 68 and standard deviation 8).

> Standardize these scores using the formula
> $$z = \frac{x - \mu}{\sigma}$$

NORMAL DISTRIBUTIONS ○○○

DON'T FORGET
Keep a table of z-values to hand. And remember that they're cumulative.

A normal distribution can be written $X \sim N(\mu, \sigma^2)$. By using the formula $Z = (Z - \mu)/\sigma$ the distribution can be standardized to $Z \sim N(0,1)$ then a curve and tables can be used to solve problems for any normally distributed variable.

> Fill in the missing items. Standardize then do a sketch graph to help you look up Φ.

If $X \sim N(50,8)$ find $P(X < 53)$, $P(X < 48)$ and $P(53 < X < 55)$.

$P(X < 53)$ $= P(Z < (53 - 50)\sqrt{8})$

$\qquad\qquad = P(Z < 1.06) = \Phi(1.06) = 0.855$

$P(X < 48)$ $= P(Z < 48 - 50)\sqrt{8})$

0 1.06

DON'T FORGET
The curve is symmetrical.

$P(53 < X < 45) =$

BINOMIAL TO NORMAL ○○○

If $X \sim B(n,p)$ and n is large so that $np > 5$ and $np(1 - p) > 5$ we can approximate the binomial using the normal.

> Fill in the formulae that convert the binomial to the normal.

Mean $\mu =$ Variance $\sigma^2 =$

If $X \sim B(48, 0.25)$ find the appropriate normal approximation to use.

> Have $n = 48$ and $p = 0.25$ so what are μ and σ^2?

CENTRAL LIMIT THEOREM ○○○

The central limit theorem (CLT) works with samples (size $n \geq 30$) and the formula is $X \sim N(\mu, \sigma^2/n)$.

> Standard error is the standard deviation of the sample mean.

If the weights of rabbits follow a normal distribution X such that $X \sim N(450, 50^2)$, and if a sample of 40 is taken, what are the mean and variance of the sample? What is the standard error?

Mean =

Variance =

Standard error =

Turn the page for some exam questions on this topic ➤

EXAM QUESTION 1

● ● ●

Brown sugar is sold in bags with masses which are normally distributed with mean 500 g and standard deviation 4 g. What proportion of bags have a weight between 499 g and 501 g?

Standardize the results, sketch a graph then use your tables.

DON'T FORGET
Finish off the question by interpreting your answer.

EXAM QUESTION 2

● ● ●

Records from a doctor's surgery show that the probability of waiting for more than 15 minutes to go into the surgery is 0.025. If the duration for waiting to go into the surgery is normally distributed with standard deviation of 2.6, what is the mean duration?

You'll need to work backwards from your tables this time. To start off, write out everything you know.

EXAM QUESTION 3

● ● ●

It is known that 2% of all batteries are faulty. What is the probability that there will be 20 or more faulty batteries in a batch of 1000?

This is a binomial model, so you'll need to approximate to a normal model.

DON'T FORGET
The binomial distribution is discrete and the normal distribution is continuous. When you standardize, you must apply the continuity correction to $X \geq 20$.

IF YOU HAVE TIME
Think what the x-value would be if the question read 'more than 20'?

Probability

Most probability questions can be solved using simple formulae but first of all you need to understand what you're being asked.

ELEMENTARY PROBABILITY ○○○

Write in words the meanings of $P(A \cap B)$, $P(A \cup B)$ and $P(A')$, then complete the addition formula.

$P(A \cap B)$

$P(A \cup B)$

$P(A')$

$P(A \cup B) =$

If $P(A) = 0.6$, $P(B) = 0.3$, $P(A \cup B) = 0.8$, find $P(A')$, $P(A \cap B)$ and $P(A' \cap B)$.

MUTUALLY EXCLUSIVE AND INDEPENDENT EVENTS ○○○

Explain mutually exclusive.

Mutually exclusive events

Complete the formula.

$P(A \cup B) =$

Explain independent events.

Independent events

Complete the formula.

$P(A \cap B) =$

EXHAUSTIVE EVENTS ○○○

Give two examples of an exhaustive event.

This is where $P(A \cup B) = 1$.

CONDITIONAL PROBABILITY ○○○

Complete the formula.

$P(A|B) =$

If there are 4 red, 3 blue and 3 yellow counters in a bag, and two are taken which are not replaced, find the probability that the second counter was red given that the first counter was blue.

Let A be taking a red counter second time and B be taking a blue, so $P(B) = 3/10$.

TREE DIAGRAMS ○○○

By showing all possible outcomes, tree diagrams make it easier to work out probabilities of combined events, especially when there is more than one way of doing something.

Construct a tree diagram for taking the counters in the above example, then work out the probability of getting a red counter and a yellow counter.

Turn the page for some exam questions on this topic ➤

EXAM QUESTION 1

●●●

In a certain town 60% of households have a freezer, 75% have a TV and 50% have both. Calculate the probability that a household chosen at random has both a TV and a freezer.

EXAM QUESTION 2

●●●

A student walks, cycles or gets a lift to school with probabilities 0.2, 0.3 and 0.5 respectively. The corresponding probabilities of being late are 0.3, 0.25 and 0.45 respectively. (a) Find the probability that the student is late on any particular day. (b) Given the student is late one day, find the probability that they cycled. (c) Given the student is not late one day, find the probability that they cycled. Answer to 3 d.p.

> Construct a tree diagram then find the probabilities.

EXAM QUESTION 3

●●●

Two darts players are throwing darts at the bullseye. The probability that the first hits the bullseye is 0.35 and the probability that the second hits the bullseye is 0.2.
(a) Find the probability that they both hit the bullseye.
(b) Find the probability that only one of them hits the bullseye.
(c) Find the probability that at least one of them hits the bullseye

Correlation

This section looks at putting a numerical value on a relationship.

PRODUCT MOMENT CORRELATION COEFFICIENT ○○○

Fill in the formula for the product moment coefficient r. Make sure you define all the terms.

Find r for this data. Find values for $\sum x_i, \sum y_i, \sum x_i^2, \sum y_i^2, x_i y_i$.
Then calculate S_{xx}, S_{yy} and S_{xy}. Use the S-values in your formula to obtain r.

x	7	12	13	15	17	18	20	25
y	20	28	23	27	36	30	31	35

DON'T FORGET
The stronger the correlation, the nearer r gets to 1 or −1 (−1 for a strong negative correlation). The weaker the correlation, the nearer r gets to 0.

Find r^2 and explain what this value means.

DON'T FORGET
The value r^2 gives the variation in percentage terms.

SPEARMAN'S RANK CORRELATION COEFFICIENT ○○○

Spearman's rank correlation coefficient r_s is a measure of how closely one set of rankings matches up with another. Write down the formula for r_s.

Complete the table, then work out the value of r_s and interpret your answer.

Name	Rank 1	Rank 2	$|d_i^2|$	d_i^2
Geoff	4	2	2	4
Harry	2	3	1	1
Ian	1	1	0	0
John	5	5	0	0
Keith	3	4	1	1

WATCH OUT
If the first ranking were to have Harry and Ian joint first place, you would split the positions and give both a ranking of 1.5.

DON'T FORGET
The product moment correlation coefficient and Spearman's rank correlation coefficient do have limitations.

Turn the page for some exam questions on this topic ➤

EXAM QUESTION 1

●●●

The weights of ten people were recorded versus shot-putting distance.

Weight (kg)	80	81	89	93	96
Putting distance (m)	14.6	13.2	14.1	14.7	15.2
Weight (kg)	100	102	110	125	76
Putting distance (m)	16.2	15.0	18.0	16.8	12.3

Calculate the product moment correlation coefficient and percentage variation. Does a heavier person throw further? Comment.

EXAM QUESTION 2

●●●

Eight people ran in two races, the 100 m and 1500 m. Here are their times. Calculate to 2 d.p. (a) Spearman's rank correlation coefficient and (b) the product moment correlation coefficient.

	A	B	C	D	E	F	G	H
100 m	11.0	14.3	12.3	12.8	16.0	15.5	11.1	17.2
1500 m	5:03	4:33	4:28	4:55	5:42	5:41	5:10	5:10

WATCH OUT
To find the product moment correlation coefficient you need to change the times for the 1500 m race into seconds only.

First make a ranking table then do the calculations.

Regression

Drawing the most accurate line of best fit on a scatter diagram is very important in order to make accurate predictions.

SCATTER DIAGRAMS

○○○

At AS-level, it is important to identify the variables, so you know which is the independent (explanatory) variable and which is the dependent (response) variable. Getting the scale right will avoid needless loss of marks.

> For this data on cars, draw a scatter diagram, stating which of the variables is the independent one

Engine size (l)	1.0	1.1	1.3	1.6	1.8	2.0	2.3	2.8	3.0	4.0
Top speed (mile h^{-1})	90	92	98	95	101	120	115	130	128	140

DON'T FORGET
The independent variable is x; in this case it's the engine size.

LEAST SQUARES REGRESSION LINE

○○○

This is a straight line with equation $y = a + bx$. Look at the engine size data. To calculate the line of regression, you need to find the values of a and b.

> Complete the steps in the working to find the equation of the regression line. Use the statistical functions on your calculator.

Step 1 First calculate S_{xy} and S_{xx}.

$S_{xx} =$

$S_{xy} =$

Step 2 Find b using the formula $b = S_{xy} / S_{xx}$.

$b =$

Step 3 Find a by substituting the mean values for x and y along with the value of a into the formula $y = a + bx$.

Mean value for $x = \dfrac{1}{n} \sum x_i =$

Mean value for $y = \dfrac{1}{n} \sum u_i =$

EXAMINER'S SECRETS
It's always a good idea to plot the line on the scatter diagram. If it doesn't fit the data, you know you've made a mistake.

WATCH OUT
Regression lines have limits. You could even estimate the top speed of a car with a 30 litre engine, which is inconceivable, or you could say a car with no engine size has top speed of 73.18 mph.

The regression line can now be used to make predictions – by using the formula or plotting the line on the graph and reading from it.

> Use your line to predict the top speed of a car with a 10 litre engine.

Turn the page for some exam questions on this topic ➤

EXAM QUESTION ● ● ●

Engine size (l)	1.0	1.1	1.3	1.6	1.8	2.0	2.3	2.8	3.0	4.0
Average Consumption mile gall^{-1}	48	53	42	40	41	35	27	29	22	19

This table shows the size of ten car engines in litres and their respective fuel consumption in miles per gallon (mile gall^{-1}).

(a) Draw a scatter graph showing this data.

(b) Calculate the least squares regression line and interpret the meaning of *a* and *b*. Add this line to your scatter graph.

(c) Use the line to estimate the average consumption of cars with 2.5 litre and 6.8 litre engines. State any problems with these estimates.

(a) Draw the graph.

WATCH OUT
Use sensible scales for the axes on your scatter diagram.

DON'T FORGET
Before you start entering all the data on your calculator, make sure you clear the memory of the last items you entered.

(b) Calculate the regression line.

Comment on your answers once you've worked them out.

(c) Estimate the average consumption for the two engine sizes.

Comment on your answers once you've worked them out.

Answer section

S<small>EE HOW YOU GOT ON BY CHECKING AGAINST</small>
<small>THE ANSWERS GIVEN HERE</small>.

Have you remembered to fill in the self-check
circles? Do this to track your progress.

For more detail on the topics covered in this book,
you can check the *Revision Express A-level Study
Guide*, your class notes or your own textbook.
You can also find exam questions and model
answers at www.revision-express.com.

Don't forget, tear out these answers and put
them in your folder for handy revision reference!

Algebra – the language of mathematics

AS AQA(A)-Me CCEA-A1 AQA(B) EDEXCEL MEI OCR WJEC-P1

You need confidence to cope with all aspects of algebra at AS-level, so give yourself lots of practice.

MANIPULATION OF FORMULAE

$y = 2x + 3$ (x)

$y - 3 = 2x$ $\therefore x = \dfrac{y-3}{2}$

$v^2 = u^2 + 2as$ (s)

$v^2 - u^2 = 2as$ $\therefore s = \dfrac{v^2 - u^2}{2a}$

$V = \pi r^2 h$ (r)

$r^2 = \dfrac{V}{\pi h}$ $\therefore r = \sqrt{\dfrac{V}{\pi h}}$

$y = \dfrac{x+1}{x-1}$ (x)

$y(x-1) = x+1$

$yx - y = x + 1$

$x(y-1) = y+1$ $\therefore x = \dfrac{y+1}{y-1}$

- Rearrange these formulae to make the letter in the bracket the subject – take it one step at a time and show all working.
- **DON'T FORGET**
  You can use the symbol $\therefore$ (therefore).

EXPANDING AND FACTORIZING

$5x(x-7)$ = (A) $5x^2 - 12x$ (B) $5x^2 + 35x$ (C) $5x^2 - 35x$

$6x^2 + 8xy$ = (A) $2x(4x+4y)$ (B) $2x(3x+4y)$ (C) $2x^2(3+4y)$

$(3x-5)(2x+7)$ = (A) $5x^2 + 11x + 2$ (B) $6x^2 + 11x - 35$ (C) $6x^2 - 11x - 35$

$2x^2 - x - 6$ = (A) $(2x+3)(x-2)$ (B) $(2x-3)(x+2)$ (C) $(2x-3)(x-2)$

$\boxed{C}$ $\boxed{B}$ $\boxed{B}$ $\boxed{A}$

- Here are some examples of taking brackets out and putting brackets in. Highlight the correct answer and write its letter in the box.

SOLVING QUADRATIC EQUATIONS

Equations of the type $ax^2 + bx + c = 0$ are called quadratic equations, and there are three ways to solve them. Get used to using all three – don't just use your favourite.

$x^2 + 3x - 10 = 0$ $(x-2)(x+5) = 0$ $x = 2$ or -5

$2y^2 + 3y - 20 = 0$ $(2y-5)(y+4) = 0$ $y = \frac{5}{2}$ or -4

$3t(t+2) + 3(t-1) = t$ $3t^2 + 6t + 3t - 3 - t = 0; 3t^2 + 8t - 3 = 0$

$(3t-1)(t+3) = 0 \therefore t = \frac{1}{3}$ or -3

$x = \dfrac{-b \pm \sqrt{b^2 - 4ac}}{2a}$

$x^2 - 3x + 1 = 0$ $a = 1, b = -3, c = 1 \therefore x = \frac{1}{2}(3 \pm \sqrt{5})$

$2x^2 + 3x - 1 = 0$ $a = 2, b = 3, c = -1 \therefore x = \frac{1}{4}(-3 \pm \sqrt{17})$

$= 0.28$ or -1.78

- Solve these equations by factorizing.
- **DON'T FORGET**
  After factorizing, check your factors are correct by multiplying out the brackets. You should obtain the expression you started with.
- Write the formula here.
- **EXAMINER'S SECRETS**
  Always quote the formula first.
- Now solve these equations,
- **EXAMINER'S SECRETS**
  In the exam you probably won't be told which method to use, so if the question asks for your answer to 2 d.p., that's a big hint to use the formula.

Solve these two equations using the formula. Leave the solutions to the first equation in surd form and give the solutions to the second equation correct to 2 d.p.

Turn the page for some exam questions on this topic ►

For more on this topic, see pages 8–9 of the *Revision Express A-level Study Guide*

EXAM QUESTION 1

Find the roots of the equation $2(x+1)(x-2) + x(x-3) = 5$, leaving your answer in surd form.

The phrase 'surd form' indicates that you should use the formula.

Multiplying out $2(x^2 - x - 2) + x^2 - 3x - 5 = 0$

$2x^2 - 2x - 4 + x^2 - 3x - 5 = 0$

Simplifying $3x^2 - 5x - 9 = 0$

Using the quadratic formula $a = 3, b = -5, c = -9$

$x = \dfrac{-b \pm \sqrt{b^2 - 4ac}}{2a}$

$\therefore x = \dfrac{5 \pm \sqrt{5^2 - (4)(3)(-9)}}{2.3} = \dfrac{5 \pm \sqrt{25 + 108}}{6} = \dfrac{5 \pm \sqrt{133}}{6}$

Note that $(4)(3)(-9) = 4.3.-9 = 4 \times 3 \times -9$

- Write the phrase in the question which suggests you need to use the formula.
- Now have a go at solving the equation.
- **DON'T FORGET**
  Quote the formula first.

EXAM QUESTION 2

(a) Expand $(1 + x)(1 + x + x^2)$ in ascending powers of x.

(b) Find as a decimal the exact value of $(1 + x)(1 + x + x^2)$ for $x = 10^{-4}$.

(a) $(1+x)(1+x+x^2) = 1 + x + x^2 + x + x^2 + x^3$

$= 1 + 2x + 2x^2 + x^3$

(b) Substitute $x = 10^{-4}$ into the answer from (a)

$\therefore (1+x)(1+x+x^2) = 1 + 2(10^{-4}) + 2(10^{-4})^2 + (10^{-4})^3$

$= 1 + 2(10^{-4}) + 2(10^{-8}) + 10^{-12}$

$= 1 + 2(0.0001) + 2(0.000\ 000\ 01) + 10^{-12}$

$= 1 + 0.0002 + 0.000\ 000\ 02 + 0.000\ 000\ 000\ 001$

$= 1.000\ 200\ 020\ 001$

- Part (a) is straightforward but take your time and show every step. For part (b), think back to your GCSE standard form.
- **DON'T FORGET**
  $a^m a^n = a^{m+n}$
  $(a^m)^n = a^{mn}$
  $a^m / a^n = a^{m-n}$
- **EXAMINER'S SECRETS**
  The phrase 'find the exact value' is a big hint that you may have do some algebra before you use your calculator.

EXAM QUESTION 3

Find the values of x which satisfy the equation

$\dfrac{3}{x+3} - \dfrac{x}{x+2} = 1$

First find common denominator

$\dfrac{3(x+2) - x(x+3)}{(x+3)(x+2)} = 1$

$3x + 6 - x^2 - 3x = 1(x^2 + 5x + 6)$

$0 = x^2 + 5x + 6 - 3x - 6 + x^2 + 3x$

$0 = 2x^2 + 5x$

$0 = x(2x+5)$

$x = 0$ or $x = -\frac{5}{2}$

- **EXAMINER'S SECRETS**
  Read the question carefully. It asks you to find the 'values', and this suggests there is more than one answer.
- To subtract fractions, first find a common denominator. To practise try $\frac{1}{2} - \frac{1}{3}$.

Algebra – quadratics

We've solved quadratic equations by formula and factorization. The third option is completing the square, a useful way to find the minimum or maximum value of a quadratic function.

COMPLETING THE SQUARE

Complete the square for these quadratic functions.

$x^2 + 4x - 7 = (x + 2)^2 - 7 - 2^2 = (x - 2)^2 - 11$

$x^2 - 10x + 3 = (x - 5)^2 + 3 - 5^2 = (x - 5)^2 - 22$

$-2x^2 - 12x - 8 = -2(x^2 + 6x + 4) = -2(x + 3)^2 + 4 - 3^2)$
$= -2[(x + 3)^2 - 5] = -2(x + 3)^2 + 10$

$4x^2 + 14x - 2 = (2x + 3\frac{1}{2})^2 - 2 - (3\frac{1}{2})^2 = (2x + 3\frac{1}{2})^2 - 14\frac{1}{4}$

Putting these quadratic functions equal to zero, we have a quadratic equation which we can now solve using our completed square.

Finish solving these equations using your answers from above. Leave your answers in surd form.

$x^2 + 4x - 7 = 0$ ∴ $(x + 2)^2 - 11 = 0$
$(x + 2)^2 = 11$ $x + 2 = \pm\sqrt{11}$ ∴ $x = -2 \pm\sqrt{11}$

$x^2 - 10x + 3 = 0$ ∴ $(x - 5)^2 - 22 = 0$
$(x - 5)^2 = 22$ $x - 5 = \pm\sqrt{22}$ ∴ $x = 5 \pm\sqrt{22}$

$-2x^2 - 12x - 8 = 0$ (divide by -2) ∴ $(2x + 3\frac{1}{2})^2 - 14\frac{1}{4} = 0$ $x^2 + 6x + 4 = 0$
$(x + 3)^2 = 5$ $x + 3 = \pm\sqrt{5}$ ∴ $x = -3 \pm\sqrt{5}$

$4x^2 + 14x - 2 = 0$ ∴ $(2x + 3\frac{1}{2})^2 - 14\frac{1}{4} = 0$
$(2x + 3\frac{1}{2})^2 = \frac{57}{4}$ $2x + \frac{7}{2} = \pm\frac{1}{2}\sqrt{57}$ ∴ $x = \frac{1}{4}(-7 \pm\sqrt{57})$

Once we've completed the square on a quadratic function we can also state the maximum or minimum value of the function by looking at the constant at the end. Just remember the smallest value for the ( )² term is zero.

Use your answers from above to find max or min values. The first one has been done for you.

$x^2 + 4x - 7 = (x + 2)^2 - 11$: min -11 when $x = -2$

$x^2 - 10x + 3 = (x - 5)^2 - 22$: min -22 when $x = 5$

$-2x^2 - 12x - 8 = -2(x + 3)^2 + 10$: max 10 when $x = -3$

$4x^2 + 14x - 2 = (2x + 3\frac{1}{2})^2 - 14\frac{1}{4}$: min $-14\frac{1}{4}$ when $x = -1\frac{3}{4}$

THE DISCRIMINANT

The discriminant $b^2 - 4ac$ tells us about the roots of a quadratic. Draw a line from each inequality to its correct root statement.

1 $b^2 - 4ac < 0$

2 $b^2 - 4ac > 0$

3 $b^2 - 4ac = 0$

3 one repeated root

1 no real roots

2 two distinct roots

Find $b^2 - 4ac$ for these quadratics and describe their roots.

$x^2 + 3x + 10 = 0$ $b^2 - 4ac = 9 - 40 = -31$ no real roots

$6 - 2x - x^2 = 0$ $b^2 - 4ac = 4 + 24 = 28$ two distinct roots

$x^2 - 20x + 25 = 0$ $b^2 - 4ac = 400 - 400 = 0$ one repeated root

Turn the page for some exam questions on this topic ▶

For more on this topic, see pages 8–9 and 10–11 of the *Revision Express A-level Study Guide*

EXAM QUESTION 1

The function $f(x) = 2x^2 + 12x - 6$ can be written in the form $p(x + q)^2 - r$, where p, q and r are positive real numbers. (a) Find the values of p, q and r. (b) Hence find coordinates of the minimum point on the graph of $f(x)$. (c) Find the coordinates of the points where the curve crosses the x-axis; leave your answers in surd form. (d) Sketch the graph.

(a) $f(x) = 2x^2 + 12x - 6 = 2(x^2 + 6x - 3) = 2[(x + 3)^2 - 3 - 9]$
$= 2[(x + 3)^2 - 12] = 2(x + 3)^2 - 24$

(b) Minimum point is $f(x) = -24$ when $x + 3 = 0$ ∴ $x = -3$
$(-3, -24)$ is the minimum point

(c) $f(x) = 0$ ∴ $2(x + 3)^2 - 24 = 0 \Rightarrow$
$2(x + 3)^2 = 24 \Rightarrow (x + 3)^2 = 12 \Rightarrow$
$x + 3 = \pm\sqrt{12}$
∴ $x = \sqrt{12} - 3$ or $-\sqrt{12} - 3$
$(\sqrt{12} = \sqrt{4}\sqrt{3} = 2\sqrt{3})$
Coordinates are
$(2\sqrt{3} - 3, 0)$ and $(-2\sqrt{3} - 3, 0)$

(d) Sketch

Part (a) is the examiner's way of asking you to complete the square. In part (b) the word 'hence' means you should use what you've already worked out.

DON'T FORGET
When completing the square on a function, you can only factor out a number (rather than divide through by the number) as the function is not yet an equation.

DON'T FORGET
If you are asked for coordinates, give the (x, y) coordinates.

EXAM QUESTION 2

(a) By substituting $y = x^{1/2}$ into the equation $4x = 13x^{1/2} - 3$, obtain a new equation in y. (b) Solve the equation to find the values of y. (c) Hence find the values of x.

(a) Have $y = x^{1/2}$, so $y^2 = x$; substituting into the equation gives $4y^2 = 13y - 3$ ∴ $4y^2 - 13y + 3 = 0$.

(b) Factorization is the quickest and simplest method to solve quadratics, so always try it first.
$(4y - 1)(y - 3) = 0$
$4y - 1 = 0$ or $y - 3 = 0$
∴ $y = 1/4$ or $y = 3$

(c) $y = 1/4$ ∴ $x^{1/2} = 1/4$ so $x = 1/16$
$y = 3$ ∴ $x^{1/2} = 3$ so $x = 9$

EXAMINER'S SECRETS
When working with fractions, it's often a lot easier to use top-heavy fractions rather than mixed numbers.

DON'T FORGET
Methods mean marks, mmm!

EXAM QUESTION 3

The curve C has the equation $y = 3x - 4 - x^2$. Show algebraically that the graph of C does not cross the x-axis.

$y = 3x - 4 - x^2 = -x^2 + 3x - 4$
∴ $a = -1, b = 3, c = -4$
$b^2 - 4ac = 9 - 4(-1)(-4) = 9 - 16 = -7$
Therefore curve C has no real roots, so its graph does not cross the x-axis.

Algebra – simultaneous equations

(AS) AQA(A)-Me CCEA-A1 AQA(B) EDEXCEL MEI OCR WJEC-P1

To find the coordinates of the point where two lines intersect, or the point(s) where a line and a curve intersect, we need to solve the equations of the lines simultaneously.

SOLVING TWO LINEAR EQUATIONS BY ELIMINATION ○○○

Find the coordinates of the point of intersection of the lines $4x + 3y = 11$ and $3x - 2y = 21$.

$4x + 3y = 11$ (1) $(1) \times 2$ $8x + 6y = 22$ (3)
$3x - 2y = 21$ (2) $(2) \times 3$ $9x - 6y = 63$ (4)

Now that the ys are the same you can add equations (3) and (4) to eliminate the ys, and find the value of x.

$(3) + (4)$ $17x = 85$ $\therefore x = 85/17 = 5$

The last part (often forgotten) is to substitute the x-value into one one of the first two equations to find the y-value.

$2x + 3y = 11$
$3y = 11 - 20 = -9$
$y = -9/3 = -3$

Therefore the point of intersection is $(5, -3)$.

DON'T FORGET
Always start by numbering your equations, then decide to get either the xs or ys the same.

Decide the multipliers to get the ys the same. Write your new equations here, number them (3) and (4)

Show the next steps here.

Substitute x = 5 into (1) to find the y-value.

EXAMINER'S SECRETS
Always finish by stating your final answer as a coordinate; you may need to do this to get your final mark.

SOLVING ONE LINEAR AND ONE QUADRATIC EQUATION BY SUBSTITUTION ○○○

Find the coordinates of the points where the line $y + 5 = x$ intersects the curve $x^2 + xy = 3$.

$y + 5 = x$ (1)
$x^2 + xy = 3$ (2)

Always look at the quadratic first (equation 2) and decide which variable it would be easiest to replace.

x ☐   y ✓

Rearrange equation (1) to get $y = x - 5$

Substitute into equation (2) $x^2 + x(x - 5) = 3$

Multiply out $x^2 + x^2 - 5x = 3$

Collect like terms and make = 0 $2x^2 - 5x - 3 = 0$

Factorize $(2x + 1)(x - 3) = 0$

Solve to obtain $x = -0.5$ and $x = 3$

Substitute $x = 3$ into (1) to give $y = -2$

Substitute $x = -0.5$ into (1) to give $y = -5.5$

So the coordinates for the points of intersection are $(3, -2)$ and $(-0.5, -5.5)$.

Tick the box to show which variable you would choose.

Follow this working and fill in the gaps.

LINKS
See solving quadratics on p. 5.

Turn the page for some exam questions on this topic ➤

For more on this topic, see pages 12–13 of the *Revision Express A-level Study Guide*

EXAM QUESTION 1 •••

The line $y = 2x - 6$ meets the x-axis at the point A. The line $2y = x + 6$ meets the y-axis at the point B. The two lines meet each other at the point C. Show that triangle ABC is isosceles.

Find the coordinates of A, B and C.

$y = 2x - 6$ (1)
$2y = x + 6$ (2)
At $A y = 0$ sub in (1)
   $0 = 2x - 6$ $\therefore x = 3$
At $B x = 0$ sub in (2)
   $2y = 6$ $\therefore y = 3$

DON'T FORGET
A sketch can often prove very useful.

At C solve (1) and (2) simultaneously
$(1) \times 2$ $2y = 4x - 12$
(2) $2y = x + 6$
$(1) - (2)$ $0 = 3x - 18$ so $3x = 18$ $\therefore x = 6$
Sub in (1) $y = 12 - 6$ $\therefore y = 6$

Therefore A is (3,0), B is (0,3) and C is (6,6).

Now show the triangle is isosceles.

$(AB)^2 = (3 - 0)^2 + (0 - 3)^2 = 9 + 9 = 18$ $\therefore AB = \sqrt{18}$
$(AC)^2 = (3 - 6)^2 + (0 - 6)^2 = 9 + 36 = 45$ $\therefore AC = \sqrt{45}$
$(BC)^2 = (0 - 6)^2 + (3 - 6)^2 = 36 + 9 = 45$ $\therefore BC = \sqrt{45}$

As $AC = BC \neq AB$ the triangle is isosceles.

LINKS
See coordinate geometry on p. 13.

EXAMINER'S SECRETS
Explain every step; you may get method marks even if you go wrong.

EXAM QUESTION 2 •••

Find the coordinates of the points of intersection of the line $x - y = 3$ with the curve $x^2 - 3xy + y^2 + 19 = 0$.

$x - y = 3$ (1)
$x^2 - 3xy + y^2 + 19 = 0$ (2)
$x = 3 + y$ (3)

Rearrange (1)
Sub (3) in (2) $(3 + y)^2 - 3y(3 + y) + y^2 + 19 = 0$
$9 + 6y + y^2 - 9y - 3y^2 + y^2 + 19 = 0$
$-y^2 - 3y + 28 = 0$
$y^2 + 3y - 28 = 0$
$(y - 4)(y + 7) = 0$
     $y = 4$ or $y = -7$

Sub $y = 4$ in (1) $x - 4 = 3$ so $x = 3 + 4$ $\therefore x = 7$
Sub $y = -7$ in (1) $x + 7 = 3$ so $x = 3 - 7$ $\therefore x = -4$

The points of intersection are $(7, 4)$ and $(-4, -7)$.

Algebra – inequalities

Inequalities can be solved using the same methods as for equations, but the answer is a range of values rather than a finite number of solutions.

○○○

LINEAR INEQUALITIES

$2x + 1 \leq 5$

$2x \leq 5 - 1$

$2x \leq 4$

$x < 2$

$8 - 5x > 23$	$2x + 1 > 4x - 7$
$-5x > 23 - 8$	$1 + 7 > 4x - 2x$
$-5x > 15$	$8 > 2x$
$x > -3$	$4 > x$
	$x > 4$

changed $\leq$ to <

divided by -5
but did not
change > to <

$4 > x$ correct
so $x < 4$

The following inequalities have been solved incorrectly. Highlight where the mistake has been made and write underneath what has gone wrong.

DON'T FORGET
When multiplying or dividing by a negative number, change the inequality sign.

THE MODULUS SIGN

○○○

Remember that the modulus sign makes everything positive, so you need to create two separate inequalities showing the positive and negative options. To solve $|2x-1| \leq 9$ consider

$(2x - 1) \leq 9$ and $-(2x - 1) \leq 9$

$2x - 1 \leq 9$ $2x - 1 \geq -9$

$2x \leq 10$ $2x \geq -8$

$x \leq 5$ $x \geq -4$

$\therefore -4 \leq x \leq 5$

Solve these two inequalities separately.

Now combine them to form one inequality.

QUADRATIC INEQUALITIES

○○○

Follow these steps every time and you shouldn't go wrong.

(1) Let quadratic = 0. (2) Solve quadratic to find critical values. (3) Sketch the graph. (4) Pick the region required. Try this one: $2x^2 + 9x - 5 \geq 0$.

Step 1 Let $2x^2 + 9x - 5 = 0$

Step 2 Factorize $(2x - 1)(x + 5) = 0$

Critical values $x = 0.5$ and $x = -5$

Step 3 Sketch the graph

EXAMINER'S SECRETS
A valid method is to sketch the graph then show its critical values and the required region.

Fill in the gaps.

Step 4 We require the region where the graph ≥ 0, i.e. the x-values for which the graph is above (or on) the x-axis.

So the solution is $x \leq -5$ and $x \geq 0.5$.

WATCH OUT
We can combine $x \geq -5$ and $x \leq 0.5$ into $-5 \leq x \leq 0.5$. We cannot combine $x \leq -5$ and $x \geq 0.5$ so we leave them separate.

Turn the page for some exam questions on this topic ▶

For more on this topic, see pages 14–15 of the *Revision Express A-level Study Guide*

EXAM QUESTION 1

○○○

Use algebra to solve $(x - 4)(x + 5) = -8$. Hence or otherwise find the set of values of x for which $(x - 4)(x + 5) \subset -8$.

First solve the equality

$x^2 + 5x - 4x - 20 + 8 = 0$

$x^2 + x - 12 = 0$

$(x - 3)(x + 4) = 0$

$x = 3$ or $x = -4$

Now solve the inequality

Need to solve
$x^2 + x - 12 < 0$
Critical values are
$x = 3$ and $x = -4$
Sketch a graph and
obtain $-4 < x < 3$

WATCH OUT
This inequality is < not $\leq$ so be careful with your answer; don't include the = part.

EXAM QUESTION 2

○○○

Sometimes inequalities can be used to solve practical problems. A farmer needs to fence a rectangular enclosure for their sheep. The length of the enclosure x is to be 8 m more than its width. The farmer has 80 m of fencing in stock, which is the maximum the farmer can use without buying some more. The farmer wants the area of the field to be greater than 240 m².

Draw a sketch, then form a linear inequality about the perimeter and solve it for x.

$x + x + (x - 8) + (x - 8) \leq 80$

$4x - 16 \leq 80$

$4x \leq 96$

$x \leq 24$

Start with a sketch showing dimensions.

x

$x - 8$

Form a quadratic inequality in x and solve it.

$x(x - 8) > 240$

$x^2 - 8x - 240 > 0$

$(x - 20)(x + 12) > 0$

Critical values are

$x = 20$ and $x = -12$

$x > 20$ or $x < -12$

-12 m is impossible

$\therefore x > 20$

Determine the set of possible values for x.

Combining (a) and (b) gives $20 < x \leq 24$

Now combine your answers to (a) and (b) and use some common sense.

Coordinate geometry – straight lines

AQA(A)·Me CCEA·A1 AQA(B) EDEXCEL MEI OCR WJEC·P1

The geometry of straight lines is simplified by a few standard formulae; understand the formulae and you're halfway there.

THE STANDARD FORMULAE

The gradient of a line is $m = \dfrac{y_2 - y_1}{x_2 - x_1}$

The gradient of a parallel line is m
The gradient of a perpendicular line is $-1/m$
There are three ways to write the equation of a line

1 $y = mx + c$ 2 $y - y_1 = m(x - x_1)$ 3 $ax + by + c = 0$

Midpoint of a line segment is at $\left(\dfrac{x_1 + x_2}{2}, \dfrac{y_1 + y_2}{2}\right)$

Distance between two points $= \sqrt{(x_2 - x_1)^2 + (y_2 - y_1)^2}$

Fill in the five formulae.

DON'T FORGET
You may want to use the points A (x_1, y_1) and $B(x_2, y_2)$ in your formulae.

FINDING THE EQUATION OF A LINE

Find the equation of the line which passes through (6,7) and (3,3).

Step 1 Find the gradient using $m = \dfrac{y_2 - y_1}{x_2 - x_1} = \dfrac{7-3}{6-3} = \dfrac{4}{3}$

Step 2 Find the line's equation using $y = mx + c$ or $y - y_1 = m(x - x_1)$

Using $y - y_1 = m(x - x_1)$ the point (3, 3) and $m = \dfrac{4}{3}$
$y - 3 = \dfrac{4}{3}(x - 3) \Rightarrow 3(y - 3) = 4(x - 3)$
$\Rightarrow 3y - 9 = 4x - 12 \Rightarrow 3y = 4x - 3$

Complete the working.

DON'T FORGET
You can use either point (6,7) or (3,3); the answer will end up the same.

Find the equation of the line which passes through the midpoint of A $(-3, 4)$ and B $(6, -2)$ and is perpendicular to the line $2y - 3x + 5 = 0$.

Step 1 Find the midpoint of A and B using
$\left(\dfrac{x_1 + x_2}{2}, \dfrac{y_1 + y_2}{2}\right) = \left(\dfrac{-3 + 6}{2}, \dfrac{4 - 2}{2}\right) = (\tfrac{3}{2}, 1)$

Step 2 Rearrange $2y - 3x + 5 = 0$ to get it in the form $y = mx + c$

$2y = 3x - 5$ so $y = \dfrac{3}{2}x - \dfrac{5}{2}$ ∴ $m = \dfrac{3}{2}$
So the gradient of the perpendicular is $-\dfrac{2}{3}$.

Complete the working.

EXAMINER'S SECRETS
It always looks impressive if you can get rid of any fractions from your equation, so multiply through by the denominator.

Step 3 Find the required equation from the midpoint and gradient

$y - y_1 = m(x - x_1) \Rightarrow y - 1 = -\dfrac{2}{3}(x - \dfrac{3}{2})$
$\Rightarrow 3(y - 1) = -2(x - \dfrac{3}{2}) \Rightarrow 3y - 3 = -2x + 3$
$\Rightarrow 3y + 2x - 6 = 0$

Use your favourite method and rearrange to get in the form $ax + by + c = 0$.

PARALLEL AND PERPENDICULAR LINES

		Parallel	Perpendicular	Neither
$4y - 2x = 11$	$2y = x + 10$	✓		
$3y - 4x = 7$	$3y + 4x = 1$		✓	
$3y + 15 = x$	$y + 3x = 12$			✓

for some quick practice, tick the box to indicate whether the pairs of lines given are parallel, perpendicular or neither.

Turn the page for some exam questions on this topic ▶

For more on this topic, see pages 36–37 of the *Revision Express A-level Study Guide*

EXAM QUESTION 1

The line L passes through the points shown. (a) Calculate the distance between A and B. (b) Find an equation for L in the form $ax + by + c = 0$, where a, b and c are integers.

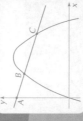

(a) A is $(-6, -6)$ and B is (2, 5) so the distance between them is
$AB^2 = (x_2 - x_1)^2 + (y_2 - y_1)^2$
$= (2 + 6)^2 + (5 + 6)^2$
$= 8^2 + 11^2 = 64 + 121 = 185$
$AB = \sqrt{185} = 13.60$ (2 d.p.)

(b) Gradient $m = \dfrac{y_2 - y_1}{x_2 - x_1} = \dfrac{5 + 6}{2 + 6} = \dfrac{11}{8}$

Equation of line $y - y_1 = m(x - x_1)$
$y + 6 = \dfrac{11}{8}(x + 6) \Rightarrow 8(y + 6) = 11(x + 6)$
$\Rightarrow 8y + 48 = 11x + 66 \Rightarrow 8y - 11x - 18 = 0$

DON'T FORGET
All you need to find the equation of a line is its gradient and a point on the line.

EXAM QUESTION 2

Here is part of the curve with equation $y = 10x - kx^2$; k is a constant. A is (0,16). The straight line L, which passes through A, B and C, is parallel to the line with equation $3y + 6x = 5$.

(a) Obtain an equation for L in the form $y = mx + c$. (b) Given the coordinates of C are (4, 8), find the value of k. (c) Calculate the coordinates of B.

EXAMINER'S SECRETS
Work your way methodically through the question one step at a time. Show all your working and quote all formulae before you use them. This shows the examiner you know the methods.

(a) L is parallel to $3y + 6x = 5$, so the gradients are the same, therefore rearrange to get $3y = -6x + 5$
$\Rightarrow y = -2x + \dfrac{5}{3}$ ∴ gradient $m = -2$. Now use the point (0,16) in $y - y_1 = m(x - x_1)$ to obtain
$y - 16 = -2(x - 0) \Rightarrow y = -2x + 16$

Take one step at a time. First find an equation for L.

(b) C has coordinates (4, 8), so substitute $x = 4$ and $y = 8$ into equation of curve
$y = 10x - kx^2$ ∴ $8 = 10(4) - k4^2 \Rightarrow 8 = 40 - 16k$
$\Rightarrow 16k = 40 - 8 = 32 \Rightarrow k = 32/16$ ∴ $k = 2$
Therefore the curve has equation $y = 10x - 2x^2$

Now use the coordinates you've been given and the equation of the curve to find the value of k.

(c) To find B, solve curve and line simultaneously
$y = -2x + 16$ (1)
$y = 10x - 2x^2$ (2)
(1) in (2): $-2x + 16 = 10x - 2x^2 \Rightarrow 2x^2 - 12x + 16 = 0$
Divide by 2: $x^2 - 6x + 8 = 0 \Rightarrow (x - 2)(x - 4) = 0$
Therefore $x = 2$ or $x = 4$; $x = 4$ at C, so $x = 2$ at B
Sub in (1): $y = -2(2) + 16 = 12$
So B has coordinates (2, 12)

B is one of the points where line L cuts the curve, so solve the curve equation and the line equation simultaneously.

LINKS
See simultaneous equations on p. 9.

Coordinate geometry – lines and curves

(AS) CCEA-A1 AQA(A) AQA(B) MEI OCR WJEC-P1 EDEXCEL-P2

Sketching and plotting curves and straight lines is a very important topic, and examiners like to test that you can do it.

STRAIGHT LINES

Any straight line can be written in the form $y = mx + c$, where m is the gradient and c is the intercept on the y-axis.

$2y = x - 4$
$y = \frac{1}{2}x - 2$
$m = \frac{1}{2}, c = -2$

$y + x = 3$
$y = 3 - x$
$m = -1, c = 3$

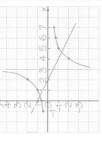

> Draw these straight lines. First plot the intercept then use the gradient to plot two more points.

GRAPH SKETCHING

To sketch a curve only requires the basic shape to be known and the coordinates where the curve crosses the axes.

$y = x \qquad y = x^2 \qquad y = x^3 \qquad y = 1/x$

> Draw graphs of these basic functions; axes have been provided for you.

GRAPH PLOTTING

To plot a curve requires points to be calculated and the curve accurately drawn on graph paper. Plot the curve with equation $y = x^3 + 2x^2 - 4$ for values of x from -3 to $+3$.

x	y
-3	-13
-2	-4
-1	-3
0	-4
1	-1
2	12
3	41

> Complete the table of values then plot the curve accurately on graph paper.

DON'T FORGET
Bigger scales mean greater accuracy.

SYLLABUS CHECK
The circle is required on the following modules: AQA(A)-P2, AQA(B)-P4, Edexcel, OCR-P3, MEI-P1, WJEC-P2.

THE CIRCLE

Equation of a circle centre (a, b) with radius r
$(x - a)^2 + (y - b)^2 = r^2$

Equation of a circle centre at $(-g, -f)$ with $r^2 = g^2 + f^2 - c$
$x^2 + y^2 + 2gx + 2fy + c = 0$

Find the equation of the circle centre $(6, 3)$ and radius 2; give your answer in the form $x^2 + y^2 + 2gx + 2fy + c = 0$

Use $(x - a)^2 + (y - b)^2 = r^2$
$(x - 6)^2 + (y - 3)^2 = 2^2$

Multiply out $x^2 - 12x + 36 + y^2 - 6y + 9 = 4$

Then simplify $x^2 + y^2 - 12x - 6y + 41 = 0$

> Write the two standard formulae for the equation of a circle.

> Fill in the gaps and complete the working.

EXAMINER'S SECRETS
Get used to both formulae. If you're not told which one to use, then go for the one that involves less work.

Turn the page for some exam questions on this topic ▶

For more on this topic, see pages 13 and 38–39 of the *Revision Express A-level Study Guide*

EXAM QUESTION 1

(a) Plot the graph of $y = \frac{2}{3 - x}$ $(x \neq 3)$ for values of x from -1 to $+7$.

(b) On the same graph draw the line with equation $2y + x = 2$.

(c) Hence give the solutions of $1 = \frac{2}{3 - x} + \frac{x}{2}$; express them to 1 d.p.

(a) Plot the graph

x	y	x	y
-1	$-\frac{1}{2}$	4	-2
0	$\frac{2}{3}$	5	-1
1	1	6	$-\frac{2}{3}$
2	2	7	$-\frac{1}{2}$
3	—		

(b) Draw the line

Rearrange $2y + x = 2$ to get $y = 1 - \frac{1}{2}x$, therefore the y-intercept is at 1 and the gradient is $-\frac{1}{2}$.

(c) Solve the equation

Rearrange $1 = \frac{2}{3 - x} + \frac{x}{2}$ to get $1 - \frac{x}{2} = \frac{2}{3 - x}$

Solutions are where curve and line intersect; these are approximately $x = 0.4$ and $x = 4.6$.

> Set up a table of values before plotting the graph.

> Start by rearranging the equation of the line into the form $y = mx + c$.

> Think about how you could rearrange the equation you've been asked to solve, and notice the question says 'hence'.

EXAM QUESTION 2

(a) A circle C has the equation $x^2 + y^2 - 4x + 2y - 4 = 0$. Find its centre and radius by rearranging the equation in the form $(x - a)^2 + (y - b)^2 = r^2$.

(b) Draw circle C accurately on graph paper.

(c) The line $y = 2x$ intersects C. Add this line to your drawing in (b).

(d) Hence find the solutions to the simultaneous equations $y = 2x$ and $x^2 + y^2 - 4x + 2y - 4 = 0$; give your answers to 1 d.p.

(a) $x^2 + y^2 - 4x + 2y - 4 = 0$
∴ $x^2 - 4x + y^2 + 2y - 4 = 0$
Complete the square
$(x - 2)^2 - 4 + (y + 1)^2 - 5 = 0$
∴ $(x - 2)^2 + (y + 1)^2 = 3^2$
Centre $(2, -1)$ radius 3

(b), (c)

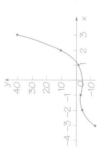

(d) Where the line intersects C gives the solutions of the equations. They can be read from the graph.
$x = 0.9$ and $y = 1.8$ or
$x = -0.9$ and $y = 1.8$

DON'T FORGET
Quote any formulae before you use them; it shows you know the methods.

> Rearrange then complete the square.

LINKS
See completing the square on p. 7.

> Use your graph to find where the line intersects the circle.

Trigonometry – radians

AQA(B) OCR-P2 CCEA-A1 AQA(A) EDEXCEL MEI WJEC-P1

Angles are usually measured in degrees. A radian is a larger unit often used in trigonometry. You need to be confident working in degrees and radians.

○○○

DEGREES AND RADIANS

Remember that $30° = \pi/6$ rad; you can build a lot of useful angles from $30°$. For example, $150° = 5 \times 30° = 5\pi/6$ rad.

Deg	360	90	540	180	30	45	15
Rad	2π	$\frac{1}{2}\pi$	3π	π	$\frac{1}{6}\pi$	$\frac{1}{4}\pi$	$\frac{1}{12}\pi$
Deg	270	210	330	300	720	450	120
Rad	$\frac{3}{2}\pi$	$\frac{7}{6}\pi$	$\frac{11}{6}\pi$	$\frac{5}{3}\pi$	4π	$\frac{5}{2}\pi$	$\frac{2}{3}\pi$

$\dfrac{\pi}{6} + \dfrac{\pi}{2}$ = $\pi(\frac{1}{6} + \frac{1}{2}) = \pi(\frac{1}{6} + \frac{3}{6})$ = $2\pi/3$

$\dfrac{\pi}{3} + \dfrac{\pi}{4}$ = $\pi(\frac{1}{3} + \frac{1}{4})$ = $\pi(\frac{4}{12} + \frac{3}{12})$ = $7\pi/12$

$\dfrac{\pi}{4} - \dfrac{\pi}{6}$ = $\pi(\frac{1}{4} - \frac{1}{6})$ = $\pi(\frac{3}{12} - \frac{2}{12})$ = $\pi/12$

$\dfrac{2\pi}{3} + \dfrac{\pi}{2}$ = $\pi(\frac{2}{3} + \frac{1}{2})$ = $\pi(\frac{4}{6} + \frac{3}{6})$ = $7\pi/6$

$2\pi - \dfrac{\pi}{6}$ = $\pi(2 - \frac{1}{6})$ = $\pi(\frac{12}{6} - \frac{1}{6})$ = $11\pi/6$

$\dfrac{3\pi}{4} + \dfrac{3\pi}{2}$ = $\pi(\frac{3}{4} + \frac{3}{2})$ = $\pi(\frac{3}{4} + \frac{6}{4})$ = $9\pi/4$

DON'T FORGET
$360° = 2\pi = 2\pi$ rad.

Complete the table by filling in the appropriate radian or degree equivalent. Leave the radian answers in terms of π.

There is no need to avoid adding and subtracting radians. Have a look at this example. Then complete the others.

DON'T FORGET
When adding or subtracting radians, take out π, treat as normal fractions then put back π at the end.

○○○

CIRCULAR MEASURE

Arc length	=	$r\theta$
Area of sector	=	$\frac{1}{2}r^2\theta$
Area of a triangle	=	$\frac{1}{2}ab\sin C$

The diagram shows a sector of a circle.

5 cm
1.5

(a) Find the arc length
(b) Find the area of the sector (2 d.p.)
(c) Find the area of the triangle (2 d.p.)
(d) Hence find the shaded area

(a) Arc length $r\theta = 5 \times 1.5 = 7.5$ cm
(b) Sector area $= \frac{1}{2}r^2\theta = \frac{1}{2}(5)^2(1.5) = 18.75$ cm^2
(c) Triangle area $= \frac{1}{2}ab\sin C = \frac{1}{2}(5)^2\sin 1.5^c = 12.47$ cm^2
(d) Shaded area $= 18.75 - 12.47 = 6.28$ cm^2

Complete these three formulae for θ measured in radians.

Complete the working and give units for your answers.

DON'T FORGET
When using $\frac{1}{2}ab\sin C$, if your angle's in radians, make sure your calculator is in radian mode.

Turn the page for some exam questions on this topic ▶

EXAM QUESTION 1

●●●

There is a straight path of length 50 m from point A to point B on a river bank. The river forms an arc of a circle centre C, radius 32 m.

A
50 m
B
32 m
C

(a) Calculate to 2 d.p. the size, in radians, of angle ACB.
(b) Calculate to 2 d.p. the length of this bend in the river.
(c) Calculate to 2 d.p. the shortest distance from C to the path.
(d) Calculate to 3 s.f. the area enclosed by the path and the river.

A
25 m
D
x
32 m
C

(a) Use $\sin x$ = opposite/hypotenuse = $25/32$
$\quad x = \sin^{-1}(25/32) = 0.897^c$
$\therefore$ angle $ACB = 2 \times 0.897 = 1.79^c$

(b) Arc length $r\theta = 32 \times 1.79 = 57.39$ m

(c) Required length is CD so use Pythagoras's theorem
$CD^2 = AC^2 - AD^2 = 32^2 - 25^2 = 399$
$CD = \sqrt{399} = 19.97$ m

(d) Shaded area = sector area – triangle area
Sector area $= \frac{1}{2}r^2\theta = \frac{1}{2} \times 32^2 \times 1.79 = 918.19$ m^2
Triangle area $= \frac{1}{2}ab\sin C = \frac{1}{2} \times 32 \times 32 \times \sin 1.79^c$
$\qquad\qquad\qquad = 499.75$ m^2
Shaded area $= 918.19 - 499.75 = 418.44$ m^2
$\qquad\qquad\qquad = 418$ m^2 (3 s.f.)

DON'T FORGET
Circular measure may involve basic trig, so recall **SohCahToa**. Sine is opposite over hypotenuse. Cosine is adjacent over hypotenuse. Tangent is opposite over adjacent.

First find angle ACB; think about an easy way to do it. Store the exact value in your calculator memory and use it in all further calculations. It makes your answers as accurate as possible.

Now find the length of the river bend.

Next find the shortest distance.

EXAM QUESTION 2

●●●

Here is a circle centre O and radius r, with chord XY. Angle XOY is 2α radians. In terms of r and α, find (a) the length of the minor arc XY and (b) the length of the chord XY. (c) If the length of the minor arc XY is $1\frac{1}{2}$ times the length of the chord XY, show that $2\alpha - 3\sin\alpha = 0$.

O
X
Y
r
2α

$0 \leq 2\alpha \leq \pi$

(a) Length of minor arc XY
Arc length $= r\theta = r.2\alpha = 2r\alpha$

(b) Length of chord XY

Let $XP = \frac{1}{2}XY$ then $\sin\alpha = XP/r$ $\therefore$ $XP = r\sin\alpha$
Hence $XY = 2r\sin\alpha$

(c) Show that $2\alpha - 3\sin\alpha = 0$

Arc $XY = 1\frac{1}{2} \times$ chord XY so $2r\alpha = \frac{3}{2} \times 2r\sin\alpha$
Therefore $2r\alpha = 3r\sin\alpha \Rightarrow 2\alpha = 3r\sin\alpha = 0$
Divide by r to get $2\alpha - 3\sin\alpha = 0$

O
α
X
P
r

Hint: split triangle OXY into two right-angled triangles.

EXAMINER'S SECRETS
If you're given a 'show that' question, make sure you show enough working to prove you've done it and not just copied the answer; the examiner won't be fooled.

Trigonometry – graphs

Graphs of trig functions help to solve trig equations. You need to know their basic shapes, their maximum and minimum values, and how often they repeat themselves (i.e. their period).

GRAPHS OF SIN X, COS X AND TAN X

$y = \sin x$ — max = 1 min = −1, period = 360°

$y = \cos x$ — max = 1 min = −1, period = 360°

$y = \tan x$ — period = 180°

> On the axes provided, sketch the graphs for sin x, cos x and tan x on the interval $0° \leq x \leq 360°$.

> Fill in the gaps giving the maximum and minimum values (where possible) and the period of each graph.

GRAPH SKETCHING

Quite often you'll be asked to sketch a graph. All you need is the basic shape and the points where the graph crosses the axes.

$y = \sin(x − 30°)$, $0° \leq x \leq 360°$ — translate 30 units right

$y = 2 \cos x$, $0° \leq x \leq 360°$ — vertical stretch scale factor 2

$y = \tan \tfrac{1}{2}x$, $−180° \leq x \leq 180°$ — horizontal stretch scale factor 2

> Use your standard graphs above to help sketch these. Describe in words how each standard graph is transformed.

LINKS See transformations of graphs on p. 45.

GRAPH PLOTTING

Accurately calculate several points then plot them on graph paper. Join the points with a smooth curve. Plot the curve $y = 1 + \sin 2x$ for $0° \leq x \leq 180°$.

> Complete the table of values for y then plot the points on graph paper and join them with a nice smooth curve.

x	0	15	30	45	60	90	105	120	135	150	180
y	1	1.5	1.87	2	1.87	1	0.5	0.13	0	0.13	1

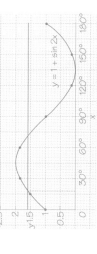

> Add the line y = 1.6 to your graph, then read off the solutions, i.e. the x-values where the curve and the line intersect.

Use your graph to find approximate solutions to $1 + \sin 2x = 1.6$.
Reading from the graph, $x = 18°$ or $x = 72°$.

DON'T FORGET The bigger the scale, the more accurate the answer.

Turn the page for some exam questions on this topic ▶

For more on this topic, see pages 42–43 of the *Revision Express A-level Study Guide*

EXAM QUESTION 1

The diagram shows part of the curve with equation $y = A + B \sin 3x$ where A and B are constants and x is in degrees. The curve passes through the points $(0°, 1)$ and $(30°, 3)$ and meets the x-axis at the points L, M and N.

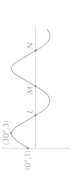

(a) Find the values of A and B.
(b) Hence determine the x-coordinates of L, M and N.

> You've been told the coordinates of two points on the curve. Use them to write some equations involving A and B.

(a) Find A and B
Substitute $(0°, 1)$ and $(30°, 3)$ into curve equation
At $(0°, 1)$ $1 = A + B\sin(3 \times 0°) = A$ ∴ $A = 1$
At $(30°, 3)$ $3 = A + B\sin(3 \times 30°) = A + B$
Substitute $A = 1$ ∴ $3 = 1 + B$ ∴ $B = 2$
Equation of curve is $y = 1 + 2\sin 3x$

(b) Find the x-coordinates for L, M, N

> Remember that you need three solutions.

At L, M and N, $y = 0$
∴ $0 = 1 + 2\sin 3x \Rightarrow 2\sin 3x = −1 \Rightarrow \sin 3x = −\tfrac{1}{2}$
$\sin^{-1}(\tfrac{1}{2}) = 30°$
∴ $3x = 210°, 330°, 570°$
$x = 70°, 110°, 190°$

EXAM QUESTION 2

(a) Sketch a graph of $y = 3\cos\tfrac{1}{2}x$ for $0° \leq x \leq 720°$. State the maximum and minimum values of y and the period of the graph.
(b) Sketch a graph of $y = 2\tan(x + 60°)$ for $−180° \leq x \leq 180°$; state the period.

> Remember to mark on the points where the graphs cross the axes.

(a) $y = 3\cos\tfrac{1}{2}x$

max = 3
min = −3
period = 720°

(b) $y = 2\tan(x + 60°)$

period = 180°

DON'T FORGET With any tan x graph you need to indicate the asymptotes (the dashed lines that the curve cannot cross).

Trigonometry – solving equations

CCEA-A1 AQA(A) AQA(B) EDEXCEL MEI OCR WJEC-P1

Trigonometry is a topic you first met at GCSE; it becomes very important in pure maths and in applied maths.

SPECIAL ANGLES

It's very useful if you can remember the ratios for particular angles. Complete the table then learn it.

Angle θ (degrees)	Angle θ (radians)	sin θ	cos θ	tan θ
0	0	0	1	0
30	π/6	1/2	√3 /2	1/ √3
45	π/4	1/ √2	1/ √2	1
60	π/3	√3 /2	1/2	√3
90	π/2	1	0	–
180	π	0	−1	0

SOLVING TRIGONOMETRIC EQUATIONS

These formulae will help you to solve trigonometric (trig) equations:

$$\sin^2 x + \cos^2 x = 1 \qquad \tan x = \frac{\sin x}{\cos x}$$

Solve $\cos(x + 30) = \frac{1}{2}$ $(0° \leq x \leq 360°)$

$$\cos^{-1}(\tfrac{1}{2}) = 60°$$
$$\therefore x + 30 = 60°, 300°$$
$$\therefore x = 30°, 270°$$

DON'T FORGET
Use a CAST (cosine, all, sine, tangent) diagram or a graph to find all the solutions.

EXAMINER'S SECRETS
Look carefully at the range of values you're given; it tells you whether your answers should be in degrees or radians.

Solve $\sin^2\theta + \cos^2\theta + 1 = 0$ $(−180° \leq \theta \leq 180°)$
From $\sin^2\theta + \cos^2\theta = 1$ we have $\sin^2\theta = 1 − \cos^2\theta$
Substitute for $\sin^2\theta$ in the equation
$$1 − \cos^2\theta + \cos\theta + 1 = 0$$
$$\cos^2\theta − \cos\theta − 2 = 0$$
$$(\cos\theta + 1)(\cos\theta − 2) = 0$$
$$\therefore \cos\theta = −1 \text{ or } \cos\theta = 2$$
$$\cos\theta = −1 \text{ gives } \theta = \pm 180°$$
$$\cos\theta = 2 \text{ gives no solution}$$

DON'T FORGET
Solving an equation like $\sin(x + 60°) = 0.5$, list all solutions for $\sin^{-1}$ 0.5 then subtract 60° from each one.

Solve $\sin\theta − \tan\theta = 0$ $(−\pi \leq \theta \leq \pi)$
Using $\tan\theta = \frac{\sin\theta}{\cos\theta}$ we have $\sin\theta − \frac{\sin\theta}{\cos\theta} = 0$
Factor out $\sin\theta$ in the equation
$$\sin\theta(1 − 1/\cos\theta) = 0$$
$$\sin\theta = 0 \text{ or } 1 − 1/\cos\theta = 0$$
$$\sin\theta = 0 \text{ or } \cos\theta = 1$$
$$\sin\theta = 0 \text{ gives } \theta = \pi, −\pi, 0; \cos\theta = 1 \text{ gives } \theta = 0$$
$$\text{so solutions are } \theta = −\pi, \pi, 0, \pi \text{ radians}$$

DON'T FORGET
Solving an equation such as tan $2\theta = 1$, you'll need to list solutions outside the range to start off with, because you'll be halving them all at the end.

Finish off these equations.

Turn the page for some exam questions on this topic ▶

For more on this topic, see pages 42–43 of the *Revision Express A-level Study Guide*

EXAM QUESTION 1

(a) Find the coordinates of the point where $y = 2\sin(2x + \pi/6)$ crosses the y-axis. (b) Find the values of x, $0 \leq x \leq 2\pi$, for which $y = \sqrt{2}$.

(a) Where the graph crosses the y-axis, $x = 0$
Substitute $x = 0$ into equation $\therefore y = 2\sin(2.0 + \pi/6)$
Hence $y = 2\sin \pi/6 = 2\cdot\frac{1}{2} = 1$ $\therefore$ coordinate is $(0,1)$

(b) Substitute $y = \sqrt{2}$ into equation
$$\sqrt{2} = 2\sin(2x + \pi/6)$$
$$\sqrt{2}/2 = \sin(2x + \pi/6)$$
now $\sqrt{2}/2 = \sqrt{2}/(\sqrt{2}\sqrt{2})$
$$= 1/\sqrt{2}$$
$$\therefore \quad 1/\sqrt{2} = \sin(2x + \pi/6)$$
now $\sin^{-1}(1/\sqrt{2}) = \pi/4$
$$2x + \pi/6 = \pi/4, 3\pi/4, 9\pi/4, 11\pi/4$$
$$(−\pi/6) \quad 2x = \pi/12, 7\pi/12, 25\pi/12, 31\pi/12$$
$$(\div 2) \qquad x = \pi/24, 7\pi/24, 25\pi/24, 31\pi/24$$

Start by asking yourself what is special about the x-coordinate of a graph when it crosses the y-axis.

DON'T FORGET
$45° = \pi/4$ rad $180° = \pi$ rad
$90° = \pi/2$ rad $360° = 2\pi$ rad

LINKS
See rationalizing surds on p. 39

EXAM QUESTION 2

Given that $\tan 15° = 2 − \sqrt{3}$, find, in the form $p + q\sqrt{3}$, where p and q are integers, the values of $\tan 75°$ and $\tan 165°$.

$$\tan 75° = \frac{1}{2 − \sqrt{3}} = \frac{1}{2 − \sqrt{3}} \cdot \frac{2 + \sqrt{3}}{2 + \sqrt{3}} = 2 + \sqrt{3}$$
$$\tan 165° = −\tan 15° = −(2 − \sqrt{3}) = \sqrt{3} − 2$$

Draw a right-angled triangle with one angle of 15°. What's the size of the third angle?

EXAM QUESTION 3

Find all angles in the range 0–360° which satisfy $2\tan x = 1 + \frac{3}{\tan x}$.

$2\tan x = 1 + 3/\tan x$ Mulitply through by tan x
So $2\tan^2 x = \tan x + 3 \Rightarrow 2\tan^2 x − \tan x − 3 = 0$
$\Rightarrow (2\tan x − 3)(\tan x + 1) = 0$
$$2\tan x − 3 = 0 \quad \text{ or } \quad \tan x + 1 = 0$$
$$\tan x = \tfrac{3}{2} \qquad\qquad \tan x = −1$$
$$\tan^{-1}(\tfrac{3}{2}) = 56.3° \qquad \tan^{-1}(−1) = −45°$$

$$x = 56.3°, 236.3°$$
$$x = 56.3°, 135°, 236.3°, 315° \qquad x = 135°, 315°$$

Polynomials – remainder theorem

AS AQA(A)=M# CCEA=A2 EDEXCEL ME1=P1 AQA(B) OCR WJEC=P2

Make sure you're confident about multiplying out brackets and factorizing quadratics and cubics. Practise on these items.

LINKS
See expanding and factorizing on p. 5.

Expand these brackets then simplify.

EXPANDING BRACKETS

$x(x+2)(x+3) = x(x^2+5x+6) = x^3+5x^2+6x$

$(x-1)^2(x+4) = (x^2-2x+1)(x+4)$
$= x^3+4x^2-2x^2-8x+x+4$
$= x^3+2x^2-7x+4$

$(2x-3)^3 = (2x-3)(4x^2-12x+9)$
$= 8x^3-24x^2+18x-12x^2+36x-27$
$= 8x^3-36x^2+54x-27$

DIVIDING POLYNOMIALS BY LONG DIVISION

Using long division, divide x^3+2x^2-x-2 by $(x-1)$

$$
\begin{array}{r}
x^2+3x+2 \\
x-1\overline{)x^3+2x^2+0x^2-x-2} \\
x^3-x^2 \\
3x^2-x \\
3x^2-3x \\
2x-2 \\
2x-2
\end{array}
$$

Answer is x^2+3x+2

Be careful, $3x^3-4x-3$ has no x^2 term, so what do you do?

Now find the remainder when $3x^3-4x-3$ is divided by x^2+2x+1

$$
\begin{array}{r}
3x-6 \\
x^2+2x+1\overline{)3x^3+0x^2-4x-3} \\
3x^3+6x^2+3x \\
-6x^2-7x-3 \\
-6x^2-12x-6 \\
5x+3
\end{array}
$$

Remainder is $5x+3$

REMAINDER THEOREM: A REMINDER

If polynomial f(x) is divided by (x − a) and the remainder is a constant, then the remainder is equal to f(a). If $f(x) = x^2 + 3x − 4$ is divided by (x − 4) then the remainder = $f(4) = 4^2 + (3 \times 4) − 4 = 24$.

Find the remainder when the polynomials are divided by the linear expressions given.

x^3+2x^2-x+5 divided by $(x-1)$
$f(1) = 1^3+2(1^2)-1+5 = 7$ remainder = 7

$2x^3-5x+16$ divided by $(x+2)$
$f(-2) = 2(-2)^3-5(-2)+16 = 10$ remainder = 10

x^3-4x^2-7 divided by $(2x-1)$
$f(\tfrac{1}{2}) = (\tfrac{1}{2})^3-4(\tfrac{1}{2})^2-7 = -7\tfrac{7}{8}$ remainder $= -7\tfrac{7}{8}$

$9x^3+3$ divided by $(3x+2)$
$f(-\tfrac{2}{3}) = 9(-\tfrac{2}{3})^3+3 = -\tfrac{8}{3}+3 = \tfrac{1}{3}$ remainder $= \tfrac{1}{3}$

SYLLABUS CHECK
Check your syllabus to see whether you need to know the remainder theorem.

Turn the page for some exam questions on this topic ▶

For more on this topic, see pages 6–7 of the *Revision Express A-level Study Guide*

EXAM QUESTION 1

(a) Expand and simplify $(x-1)(2x+3)(4-x)$ arranging your answer in descending powers of x.
(b) Find the remainder when the answer to (a) is divided by x^2+x+1.

(a) First expand and simplify

$(x-1)(2x+3) = 2x^2+3x-2x-3 = 2x^2+x-3$

$(2x^2+x-3)(4-x) = 8x^2+4x-12-2x^3-x^2+3x$
$= -2x^3+7x^2+7x-12$

(b) Now use long division

$$
\begin{array}{r}
-2x+9 \\
x^2+x+1\overline{)-2x^3+7x^2+7x-12} \\
-2x^3-2x^2-2x \\
9x^2+9x-12 \\
9x^2+9x+9 \\
-21
\end{array}
$$

Remainder is −21

EXAMINER'S SECRETS
Double-check this as you're going to use it in the next part – it's worth the effort.

EXAM QUESTION 2

The remainder obtained when $3x^3-6x^2+ax-1$ is divided by $x+1$ is equal to the remainder when the same expression is divided by $x-3$. Find the value of a.

Let $f(x) = 3x^3-6x^2+ax-1$
$\therefore f(-1) = 3(-1)^3-6(-1)^2+a(-1)-1$
$= -3-6-a-1 = -10-a$
Also $f(3) = 3(3)^3-6(3)^2+a(3)-1$
$= 81-54+3a-1 = 26+3a$
Remainders are the same, so $-10-a = 26+3a$
$4a = -36$
$a = -9$

EXAM QUESTION 3

The expression $6x^2+x+7$ leaves exactly the same remainder when it is divided by $x-a$ and $x+2a$ ($a\neq0$). Calculate the value of a.

Let $f(x) = 6x^2+x+7$ $\therefore f(a) = 6a^2+a+7$
and $f(-2a) = 6(-2a)^2+(-2a)+7$
$= 24a^2-2a+7$
Remainders are the same, so
$6a^2+a+7 = 24a^2-2a+7$
$0 = 18a^2-3a$
$0 = 3a(6a-1)$
$\therefore$ either $3a = 0$ or $6a-1 = 0$
$a = 0$ $6a = 1$
$a = 0$ is not possible, so $a = \tfrac{1}{6}$

WATCH OUT
Look closely at a question to see if you're given any conditions for your unknown number. In this question you're told $a\neq0$.

Polynomials – factor theorem

Following on from the remainder theorem comes the factor theorem. We can use it to find roots of polynomials and hence sketch their graphs.

FACTOR THEOREM

If $(x - a)$ is a factor of a polynomial $f(x)$ then $f(a) = 0$ (i.e. there is no remainder). So to factorize the cubic $f(x) = x^3 - 2x^2 - 5x + 6$ we follow these three steps.

Step 1 Use the remainder theorem, and trial and improvement, to find a factor. Notice that the constant at the end is a 6, so it's worth trying a factor of 6, e.g. 1, 2 or 3. Try $x = 1$ so that $f(1)$ is

$(1)^3 - 2(1)^2 - 5 \times 1 + 6 = 0$
$(x - 1)$ is a factor

Do the working and state the factor.

Step 2 Use long division, or a similar method, to divide the cubic by its factor and find the quadratic.

$$x - 1 \overline{\smash{)}\begin{array}{l} x^2 - x - 6 \\ x^3 - 2x^2 - 5x + 6 \\ \underline{x^3 - x^2} \\ -x^2 - 5x \\ \underline{-x^2 + x} \\ -6x + 6 \\ \underline{-6x + 6} \end{array}}$$

Show your method here.

Step 3 Factorize the quadratic to give you the other two factors then write the factorized expression for $f(x)$.

$x^2 - x - 6 = (x - 3)(x + 2)$
$f(x) = (x - 1)(x - 3)(x + 2)$

Factorize the quadratic to find the other two factors; then write $f(x)$ in factors.

SKETCHING GRAPHS OF POLYNOMIALS

To sketch the graph of a polynomial, first use the factor theorem to find the factors. Then, by equating the polynomial to zero, find its roots and mark them on the sketch. Finally decide in between two of the roots whether $f(x)$ is positive or negative. Now you're ready to sketch the graph.

If $f(x) = 2x^3 + 7x^2 - 7x - 12$ factorizes to $f(x) = (2x - 3)(x + 4)(x + 1)$ then solving $f(x) = 0$ gives $(2x - 3)(x + 4)(x + 1) = 0$; $x = \frac{3}{2}, -4, -1$.

Mark $f(0)$ on the axis then sketch the curve.

Choose a value between two of the roots, e.g. $x = 0$
Here $f(0) = -12$

[graph sketch with points -4, -1, $\frac{3}{2}$, and -12]

DON'T FORGET
Finding $f(0)$ is a good idea anyway, as on your sketch you need to mark the point where the graph crosses the y-axis as well as the x-axis.

Turn the page for some exam questions on this topic ▶

EXAM QUESTION 1

(a) Show that $2x + 1$ is a factor of $f(x) = 2x^3 + x^2 - 8x - 4$.
(b) Hence find the values of x for which $f(x) = 0$.

(a) If $2x + 1$ is a factor then $f(-\frac{1}{2}) = 0$
$f(-\frac{1}{2}) = 2(-\frac{1}{2})^3 + (-\frac{1}{2})^2 - 8(-\frac{1}{2}) - 4$
$= -\frac{1}{4} + \frac{1}{4} + 4 - 4 = 0$

(b) $$2x + 1 \overline{\smash{)}\begin{array}{l} x^2 \qquad -4 \\ 2x^3 + x^2 - 8x - 4 \\ \underline{2x^3 + x^2} \\ 0 - 8x - 4 \\ \underline{-8x - 4} \end{array}}$$

$f(x) = (2x + 1)(x^2 - 4) = (2x + 1)(x - 2)(x + 2)$
So when $f(x) = 0$ we have $(2x + 1)(x - 2)(x + 2) = 0$
Therefore $x = -\frac{1}{2}, x = 2, x = -2$

DON'T FORGET
To factorize $x^2 - y^2$ use the difference of two squares, so $x^2 - y^2 = (x + y)(x - y)$.

EXAM QUESTION 2

Given that $f(x) = x^3 + x^2 - 5x - 2$, (a) show that $(x - 2)$ is a factor of $f(x)$; (b) factorize $f(x)$ completely and find the exact values for which $f(x) = 0$; (c) hence, by substituting $x = \tan t$, solve the equation $\tan^3 t + \tan^2 t - 5\tan t - 2 = 0$ ($0 \le t \le 360°$); give the solutions to 1 d.p.

(a) If $x - 2$ is a factor, then $f(2)$ should equal zero
$\therefore f(2) = (2)^3 + (2)^2 - 5(2) - 2 = 8 + 4 - 10 - 2 = 0$

$$x - 2 \overline{\smash{)}\begin{array}{l} x^2 + 3x + 1 \\ x^3 + x^2 - 5x - 2 \\ \underline{x^3 - 2x^2} \\ 3x^2 - 5x \\ \underline{3x^2 - 6x} \\ x - 2 \\ \underline{x - 2} \end{array}}$$

(b) $f(x) = (x - 2)(x^2 + 3x + 1)$
Using the formula on $x^2 + 3x + 1 = 0$ gives
$x = \dfrac{-b \pm \sqrt{b^2 - 4ac}}{2a} = \dfrac{-3 \pm \sqrt{9 - 4}}{2} = \dfrac{-3 \pm \sqrt{5}}{2}$

Exact solutions are $x = 2$, $\frac{1}{2}(-3 + \sqrt5)$, $\frac{1}{2}(-3 - \sqrt5)$

(c) Sub $x = \tan t$ into $\tan^3 t + \tan^2 t - 5\tan t - 2 = 0$
to get $x^3 + x^2 - 5x - 2 = 0$ then use answers in (b) to get
$x = 2$, $x = -0.38$, $x = -2.62$
$\tan t = 2$ $\tan t = -0.38$ $\tan t = -2.62$
$\tan^{-1} 2 = 63.4°$ $\tan^{-1} -0.38 = -20.9°$ $\tan^{-1} -2.62 = -69.1°$
So $t = 63.4°, 110.9°, 159.1°, 243.4°, 290.9°, 339.1°$

[three angle diagrams showing $63.4°$, $20.9°$, $69.1°$]

EXAMINER'S SECRETS
The phrase 'find the exact value' means your answer is either a whole number or it must be left as a fraction or surd. This is a big hint for part (b).

LINKS
See solving quadratics on p. 5; see trigonometric equations on p. 21.

Polynomials – binomial theorem

Expanding a binomial is relatively easy as long as you can remember either Pascal's triangle (if the power is small) or the expansion formula (if the power is large).

PASCAL'S TRIANGLE

Complete Pascal's triangle to show the coefficients when $(a + b)$ is expanded up to the power of 6.

$(a+b)^1$
$(a+b)^2$
$(a+b)^3$
$(a+b)^4$
$(a+b)^5$
$(a+b)^6$

```
            1   1
          1   2   1
        1   3   3   1
      1   4   6   4   1
    1   5  10  10   5   1
  1   6  15  20  15   6   1
```

First expand $(a + b)^5$.

Use Pascal's triangle to expand $(2 - x)^5$ in increasing powers of x.
$(a + b)^5 = a^5 + 5a^4b + 10a^3b^2 + 10a^2b^3 + 5ab^4 + 1b^5$

Write out the expansion of $(2 - x)^5$ and simplify each term as far as possible.

Now compare $(a + b)^5$ with $(2 - x)^5$ and substitute $a = 2$ and $b = -x$.
$(2 - x)^5 = 1.2^5 + 5.2^4(-x) + 10.2^3(-x)^2$
$\qquad + 10.2^2(-x)^3 + 5.2^1(-x)^4 + 1(-x)^5$
$= 32 - 80x + 80x^2 - 40x^3 + 10x^4 - x^5$

BINOMIAL THEOREM

There are two versions of the binomial expansion; practise both.

Complete the expansions for $(a + x)^n$ up to the term in x^3.

$(a + x)^n = a^n + {}^nC_1a^{n-1}x + {}^nC_2a^{n-2}x^2 + {}^nC_3a^{n-3}x^3 + \ldots$

$(a + x)^n = a^n + n.a^{n-1}x +$
$\dfrac{n(n-1)}{2!}a^{n-2}x^2 + \dfrac{n(n-1)(n-2)}{3!}a^{n-3}x^3 + \ldots$

${}^nC_r = \dbinom{n}{r} = \dfrac{n!}{r!(n-r)!}$

DON'T FORGET
$n! = n(n-1)(n-3)\ldots3.2.1$

Use the first expansion and simplify your answer.

Expand $(1 + 2x)^{12}$ in ascending powers of x up to and including the term in x^4. Hence evaluate $(1.02)^{12}$ correct to 5 d.p.
$(1 + 2x)^{12} = 1^{12} + {}^{12}C_1(2x) + {}^{12}C_2(2x)^2$
$\qquad + {}^{12}C_3(2x)^3 + {}^{12}C_4(2x)^4 + \ldots$
$= 1 + 24x + 264x^2 + 1760x^3 + 7920x^4 + \ldots$

Substitute x = 0.01 into the expansion then evaluate your answer.

Comparing $(1.02)^{12}$ with $(1 + 2x)^{12}$ then $1.02 = 1 + 2x$ and $x = 0.01$
$(1.02)^{12} = 1 + 24(0.01) + 264(0.01)^2$
$\qquad + 1760(0.01)^3 + 7920(0.01)^4$
$= 1.268\ 2392 = 1.268\ 24$ (5 d.p.)

Write the expansion up to the x^2 term.

The coefficient of x^2 in the expansion of $(1 + \frac{1}{2}x)^n$ is 7. Find n.
$(1 + \frac{1}{2}x)^n = 1^n + n1^{n-1}(\frac{1}{2}x) + [n(n-1)]1^{n-2}(\frac{1}{2}x)^2]/2!$

Simplify and solve to find n.

We're told the coefficient of x^2 is 7, so $7 = n(n-1)$
$7 \times 2! \times 4 = n(n-1) \therefore 56 = n(n-1)$
$56 = n(n-1)$
$n = 8$ as $8 \times 7 = 56$

Turn the page for some exam questions on this topic ▶

For more on this topic, see pages 32–33 of the *Revision Express A-level Study Guide*

EXAM QUESTION 1

(a) Use the binomial theorem to expand $(2 + 10x)^4$, giving each coefficient as an integer. (b) Use your expansion to find the exact value of $(1002)^4$ stating the value of x which you have used.

(a) It's easiest here to use Pascal's triangle as the bracket is only to the power of 4
Coefficients are 1, 4, 6, 4, 1, so $(2 + 10x)^4$ is $1.2^4 + 4.2^3.10x + 6.2^2.(10x)^2 + 4.2^1.(10x)^3 + 1.(10x)^4$
$= 16 + 320x + 2400x^2 + 8000x^3 + 10000x^4$

An alternative method would be to factor out a 2 to get $2^4(1 + 5x)^4$ then expand $(1 + 5x)^4$, but don't forget to multiply the expansion by 2^4 at the end.

(b) Compare $(2 + 10x)^4$ with $(1002)^4$
$\therefore 2 + 10x = 1002 \Rightarrow 10x = 1000 \Rightarrow x = 100$
Substitute x = 100 into expansion
$(1002)^4 = 16 + 320(100) + 2400(100)^2 + 8000(100)^3 +$
$\qquad\qquad 10000(100)^4$
$= 16 + 32\,000 + 24\,000\,000$
$\qquad + 8\,000\,000\,000 + 1\,000\,000\,000\,000$
$= 1\,008\,024\,032\,016$

EXAM QUESTION 2

(a) Find the non-zero values of a and b in the expansion of $(a + x/b)^8$ in ascending powers of x, when the first term is 256 and the coefficient of x^3 is three times the coefficient of x^2.
(b) Using your values of a and b, give the first four terms in the binomial expansion of $(a + x/b)^8$ in descending powers of x.

EXAMINER'S SECRETS
Follow the instructions; give ascending powers of x when asked for ascending powers and descending powers when asked for descending powers.

(a) Only the first four terms of the expansion are needed; why is this?

$(a + \frac{1}{b}x)^8 = a^8 + {}^8C_1a^7(\frac{1}{b}x) + {}^8C_2a^6(\frac{1}{b}x)^2 + {}^8C_3a^5(\frac{1}{b}x)^3$
$= a^8 + 8a^7(\frac{1}{b}x) + 28a^6(\frac{1}{b}x)^2 + 56a^5(\frac{1}{b}x)^3$

We're told the first term is 256 $\therefore a^8 = 256 \Rightarrow a = 2$
We're told coeff. of x^3 is three times coeff. of x^3
$\therefore 28a^6/b^3 = 3 \times 56a^5/b^2$
$28a^6b^3 = 3 \times 56a^5b^2$ (cross-multiplying)
$28ab = 168$ (dividing by a^5b^2)
$ab = 6$
Substituting a = 2 gives $2b = 6 \therefore b = 3$

(b) Descending powers of x needed, so expand
$(\frac{1}{3}x + 2)^8 = (\frac{1}{3}x)^8 + {}^8C_1(\frac{1}{3}x)^7.2 + {}^8C_2(\frac{1}{3}x)^6.2^2 + {}^8C_3(\frac{1}{3}x)^5.2^3$
$= \frac{1}{6561}x^8 + \frac{16}{2187}x^7 + \frac{112}{729}x^6 + \frac{448}{243}x^5$

Differentiation

Differentiation is for calculating the **gradient of the tangent to a curve** at any point on the curve. If $y = f(x)$ then we can write dy/dx as $f'(x)$ or simply y'; they all mean the same thing.

WHAT YOU SHOULD KNOW ○○○

If the gradient of a tangent to a curve is m, then the gradient of the normal is what?

    (A) m^2 (B) $1/m$ (C) $-1/m$

To find a maximum or minimum point on the curve we find dy/dx and set it equal to what?

    (A) x   (B) 0   (C) 1

To find the values of x for which $f(x)$ is an increasing function, state the condition we solve for $f'(x)$.

    (A) $f'(x) = 0$ (B) $f'(x) < 0$ (C) $f'(x) > 0$

To find the values of x for which $f(x)$ is a decreasing function, state the condition we solve for $f'(x)$.

    (A) $f'(x) = 0$ (B) $f'(x) < 0$ (C) $f'(x) > 0$

What is the result of differentiating $y = kx^n$?

    (A) $dy/dx = knx^{n-1}$ (B) $dy/dx = \frac{k}{n}x^{n-1}$ (C) $dy/dx = kx^{n+1}/(n+1)$

Highlight the correct answer – A, B or C.

USE OF DIFFERENTIATION WITH CURVE SKETCHING ○○○

Differentiating to find maxima and minima can help when sketching graphs of functions. Sketch the graph of $y = 2x^3 + 3x^2 - 12x + 1$.

Step 1 Find, when possible, the coordinates where the curve crosses the axes.

when $x = 0$, $y = 2(0)^3 + 3(0)^2 - 12(0) + 1 = 1$
when $y = 0$, $0 = 2x^3 + 3x^2 - 12x + 1$ (no obvious solutions)

Put $x = 0$ then $y = 0$ into the equation of the curve to find where it crosses the axes.

Step 2 Find any turning points by differentiating.

$y = 2x^3 + 3x^2 - 12x + 1$ ∴ $dy/dx = 6x^2 + 6x - 12$
∴ $6x^2 + 6x - 12 = 0 \Rightarrow x^2 + x - 2 = 0$
$\Rightarrow (x + 2)(x - 1) = 0$ ∴ $x = -2$ or $x = 1$

Substitute the x-values into the curve equation
when $x = -2$, $y = 2(-2)^3 + 3(-2)^2 - 12(-2) + 1 = 21$
when $x = 1$, $y = 2(1)^3 + 3(1)^2 - 12(1) + 1 = -6$

Differentiate then put $dy/dx = 0$; solve for x.

DON'T FORGET
Once you've found the x-coordinates of any turning points, find their corresponding y-values.

Step 3 Decide whether the turning points are maxima, minima or points of inflexion.

x	-3	-2	-1
dy/dx	+	0	−

x	0	1	2
dy/dx	−	0	+

$(1, -6)$ is a minimum ·
$(-2, 21)$ is a maximum

Examine the sign of dy/dx before and after the turning point to determine its nature.

Finally sketch the curve.

Turn the page for some exam questions on this topic ▶

For more on this topic, see pages pages 48–49 and 50–51 of the *Revision Express A-level Study Guide*

EXAM QUESTION 1

(a) Find the coordinates of any stationary points on the curve $f(x) = 5x^6 - 12x^5$ and distinguish between them.
(b) Hence sketch the curve and give the values of x for which $f(x)$ is an increasing function.

$f(x) = 5x^6 - 12x^5$ ∴ $f'(x) = 30x^5 - 60x^4$
at turning points $f'(x) = 0$
  ∴ $30x^5 - 60x^4 = 0$
    $x^4(x - 2) = 0$
$x^4 = 0$ or $x - 2 = 0$
$x = 0$ or $x = 2$
$f(0) = 5(0)^6 - 12(0)^5 = 0$
$f(2) = 5(2)^6 - 12(2)^5 = -64$

x	-1	0	1
$f'(x)$	−	0	−

x	1	2	3
$f'(x)$	−	0	+

$(0, 0)$ point of inflexion $(2, -64)$ minimum

Increasing function is when $f'(x) > 0$, i.e. when gradient is positive, so $x > 2$

DON'T FORGET
$f'(x) > 0$ for an increasing function.

EXAM QUESTION 2

An open metal tank with a square base is made from 16 m^2 of sheet metal. Find the tank dimensions needed to maximize the volume and find this maximum volume.

Let the base length be x and the height be h
Volume $= x^2h$
Area $= x^2 + 4xh$
∴ $16 = x^2 + 4xh$
$h = (16 - x^2)/4x$
Sub into $V = x^2h$ to get
$V = x^2(16 - x^2)/4x = 4x - x^3/4$
For max volume $dV/dx = 0$
∴ $4 - 3x^2/4 = 0$
$x = \pm 4/\sqrt{3}$
$x = -4/\sqrt{3}$ isn't a possible length, so $x = 4/\sqrt{3}$ m
Sub $x = 4/\sqrt{3}$ in $h = (16 - x^2)/4x$ gives $h = 2/\sqrt{3}$
∴ $V = x^2h = \left(\frac{4}{\sqrt{3}}\right)^2 \left(\frac{2}{\sqrt{3}}\right) = \frac{32}{3\sqrt{3}} = \frac{32\sqrt{3}}{9}$ m^3

DON'T FORGET
Differentiation is often used for finding the answers to practical problems.

DON'T FORGET
Start with a sketch showing the letters you're using for the dimensions of the tank, then use the information you're given: area is 16 m^2 and the volume is to be found.

DON'T FORGET
Check it's a maximum by examining the sign of dy/dx.

Integration

Integration is the reverse of differentiation – given the gradient function of a curve (and a point on the curve) we can integrate to obtain the equation of the curve.

THE REVERSE OF DIFFERENTIATION

Link each gradient function with the correct curve equation; use the values for x and y to work out the constant.

DON'T FORGET
If $\dfrac{dy}{dx} = kx^n$ then
$y = \dfrac{kx^{n+1}}{n+1} + c \ (n \neq -1)$.

1	$dy/dx = 2x + 5$	$x = 1, y = 0$	$y = \frac{1}{7}x^3 - 2x + 2$
2	$dy/dx = x^2 - 2$	$x = 3, y = 5$	$y = 2\sqrt{x} + 3x - 6$
3	$dy/dx = x^4 - 3x$	$x = 0, y = 2$	$y = x^2 + 5x + 1$
4	$dy/dx = 2x + 5$	$x = 0, y = 1$	$y = x^2 + 5x - 6$
5	$dy/dx = 1/\sqrt{x} + 3$	$x = 4, y = 10$	$y = \frac{1}{5}x\sqrt{x} + x^2 - 11\frac{1}{5}$
6	$dy/dx = \sqrt{x} + 2x$	$x = 4, y = 10$	$y = \frac{1}{5}x^5 - \frac{3}{2}x^2 + 2$

(answers) ② ⑤ ④ ① ⑥ ③

THE AREA UNDER A CURVE

The area under a curve can be found using integration. If we require the area between a curve and the x-axis, we use $\int y\,dx$ and if we require the area between a curve and the y-axis we use $\int x\,dy$. To find the area in the first quadrant between the curve $y^2 = x$, the y-axis and the line $y = 3$ follow these steps.

Step 1 Find where the curve intersects the axes then sketch the curve, add the line $y = 3$ and shade the required area.
When $x = 0, y = 0$

Complete step 1.

Step 2 Decide whether $\int y\,dx$ or $\int x\,dy$ is needed.
We need $\int x\,dy$ so the area $= \int_0^3 y^2\,dy$

Complete step 2.

Step 3 Integrate and substitute the limits.
Area $= [\frac{1}{3}y^3]_0^3 = \frac{1}{3}(3^3) - \frac{1}{3}(0^3) = 9$ unit^2

Complete step 3.

Follow the same three steps to find the area between the curve $y = 2x - x^2$ and the x-axis from $x = 0$ to $x = 3$.

Complete the three steps again, but think carefully how to find the area required.

Step 1
$y = 2x - x^2 \ \therefore \ y = x(2 - x)$
when $y = 0, x = 0$ or $x = 2$

EXAMINER'S SECRETS
Always show you've substituted in both limits, even if you can see that one is zero. It tells the examiner you know what you're doing.

Step 2
Area $A = \int_0^2 y\,dx = \int_0^2 (2x - x^2)\,dx$
Area $B = \int_2^3 y\,dx = \int_2^3 (2x - x^2)\,dx$

DON'T FORGET
The area lying above the axis will have a positive value and the area lying below will have a negative value, so be careful.

Step 3
Area $A = [x^2 - \frac{1}{3}x^3]_0^2$ Area $B = [x^2 - \frac{1}{3}x^3]_2^3$
$A = 2^2 - \frac{1}{3}\cdot2^3 - (0 - 0) = 1\frac{1}{3}$
$B = 3^2 - \frac{1}{3}\cdot3^3 - (2^2 - \frac{1}{3}\cdot2^3) = -1\frac{1}{3}$
Area required $= 1\frac{1}{3} + 1\frac{1}{3} = 2\frac{2}{3}$ unit^2

The negative sign on area B indicates that it lies below the x-axis; remove the negative sign and add to area A.

Turn the page for some exam questions on this topic ▶

For more on this topic, see pages 56–57 and 60–61 of the *Revision Express A-level Study Guide*

EXAM QUESTION 1

Find the area bounded by $y^2 = x + 4$, the y-axis and the line $y = 4$.

when $x = 0, y^2 = 4 \ \therefore \ y = \pm 2$
when $y = 0, x = -4$
Area $= \int_{-2}^{4} x\,dy = \int_{-2}^{4} (y^2 - 4)\,dy$
$= [\frac{1}{3}y^3 - 4y]_{-2}^{4} = 10\frac{2}{3}$ unit^2

EXAM QUESTION 2

Find the area of the region enclosed between the line $y = 2$ and the curve $y = 5x - 4 - x^2$.

First find where the line and curve intersect then do a sketch.

Solve $y = 2$ (1) $\ y = 5x - 4 - x^2$ (2)
sub (1) in (2) $2 = 5x - 4 - x^2 \Rightarrow x^2 - 5x + 6 = 0$
$\Rightarrow (x - 3)(x - 2) = 0 \Rightarrow x = 3$ or 2

$\text{Area} = \int_2^3 (5x - 4 - x^2)\,dx$
$= [\frac{5}{2}x^2 - 4x - \frac{1}{3}x^3]_2^3$
$= \frac{5}{2}(3^2) - 4.3 - \frac{1}{3}(3^3)$
$\quad - [\frac{5}{2}(2^2) - 4.2 - \frac{1}{3}(2^3)]$
$= 2\frac{1}{6}$ unit^2

Area of rectangle $= 2$ unit^2
Shaded area $= 2\frac{1}{6} - 2 = \frac{1}{6}$ unit^2

WATCH OUT
Don't try to do too much in your head – write it all down. It might take longer but you're less likely to make a mistake.

EXAM QUESTION 3

Find the area bounded by the curves $y = x^2 - 4x$ and $y = 6x - x^2$.

Find where the two curves intersect then do a sketch.

Solve $y = x^2 - 4x$ (1) $\quad y = 6x - x^2$ (2)
sub (1) in (2) $x^2 - 4x = 6x - x^2 \Rightarrow 2x^2 - 10x = 0$
$\Rightarrow 2x(x - 5) = 0 \Rightarrow x = 0$ or 5
$y = x^2 - 4x = x(x - 4)$
when $y = 0, x = 0$ or 4
$y = 6x - x^2 = x(6 - x)$
when $y = 0, x = 0$ or 6
Required area $= A + B$

$A = \int_0^5 (6x - x^2)\,dx - \int_4^5 (x^2 - 4x)\,dx$
$= [3x^2 - \frac{1}{3}x^3]_0^5 - [\frac{1}{3}x^3 - 2x^2]_4^5 = 31$ unit^2
$B = \int_0^4 (x^2 - 4x)\,dx = [\frac{1}{3}x^3 - 2x^2]_0^4 = -10\frac{2}{3}$ unit^2
Required area $= 31 + 10\frac{2}{3} = 41\frac{2}{3}$ unit^2

Integration – trapezium rule

○○○

The trapezium can be used to find an approximate value for the area under a curve, without having to integrate.

USING THE TRAPEZIUM RULE

If the area is divided into n strips, each of width h, then the formula is

$$\int f(x)\,dx \approx \tfrac{1}{2}h\{y_0 + y_n + 2(y_1 + y_2 + K + y_{n-1})\}$$

Find an approximate value for $\int_0^1 (1+x)^{-1}\,dx$ with five strips.

Since x goes from 0 to 1, then with five strips h would be $\tfrac{1}{5}$.

x	0	$\frac{1}{5}$	$\frac{2}{5}$	$\frac{3}{5}$	$\frac{4}{5}$	1
y	1	$\frac{5}{6}$	$\frac{5}{7}$	$\frac{5}{8}$	$\frac{5}{9}$	$\frac{1}{2}$

$$\therefore \int_0^1 (1+x)^{-1}\,dx = \tfrac{1}{2\cdot5}\{1 + \tfrac{1}{2} + 2(\tfrac{5}{6} + \tfrac{5}{7} + \tfrac{5}{8} + \tfrac{5}{9})\} = 0.6956$$

Fill in the formula for the trapezium rule.

DON'T FORGET
Five strips means there are six ordinates, i.e. there are six values for x and y.

Complete the table of values for y.

Complete the calculation.

ACCURACY OF THE TRAPEZIUM RULE

○○○

The more strips or ordinates used, the more accurate the answer.

If $A = \int_1^2 \dfrac{3x^4 + 2}{x^2}\,dx$

(a) Find A by integration. (b) Use the trapezium rule to find an approximate value for A by (i) taking 6 ordinates and (ii) taking 11 ordinates. (c) Calculate the percentage error in using the trapezium rule, giving your answer correct to 2 d.p.

$$A = \int_1^2 \frac{3x^4 + 2}{x^2}\,dx + \int_1^2 (3x^2 + 2x^{-2})\,dx = [x^3 - 2x^{-1}]_1^2$$
$$= (2^5 - \tfrac{2}{2}) - (1^3 - \tfrac{2}{1}) = 7 - (-1) = 8 \text{ unit}^2$$

x goes from 1 to 2 so h = 0.2 (6 ordinates).

x	1	1.2	1.4	1.6	1.8	2
y	5	5.709	6.900	8.461	10.337	12.5

$A = \tfrac{1}{2}(0.2)\{5 + 12.5 + 2(5.709 + 6.900 + 8.461 + 10.337)\}$
$= 8.0315662$

x goes from 1 to 2 and h = 0.1 (11 ordinates).

x	1	1.1	1.2	1.3	1.4	1.5
y	5	5.283	5.709	6.253	6.900	7.639

x	1.6	1.7	1.8	1.9	2
y	8.461	9.362	10.337	11.384	12.5

$A = \tfrac{1}{2}(0.1)\{5 + 12.5 + 2(5.283 + 5.709 + 6.253 + 6.900$
$+ 7.639 + 8.461 + 9.362 + 10.337 + 11.384)\}$
$= 8.0079103$

% error = 100 × (8.031 5662 − 8)/8 = 0.39% (2 d.p.)
% error = 100 × (8.007 9103 − 8)/8 = 0.10% (2 d.p.)

DON'T FORGET
Six ordinates means 5 strips, and 11 ordinates means 10 strips.

Integrate this by separating it into two fractions and simplifying.

Complete the tables, giving the y-values correct to 3 d.p. then substitute into the trapezium rule and evaluate.

EXAMINER'S SECRETS
Use the memory on your calculator to store the accurate values (not the rounded values) then your final answer will be as accurate as possible.

DON'T FORGET
% error = $\dfrac{\text{actual error}}{\text{actual value}}$ ×100%

Calculate the percentage error.

Turn the page for some exam questions on this topic ►

EXAM QUESTION 1

●●●

Find an approximate value for $\int_0^{\pi/2} \sin x\,dx$ by using the trapezium rule with eight strips. Give your answer correct to 5 d.p.

Eight strips means 9 ordinates so h = π/16

x	0	π/16	π/8	3π/16	π/4
y	0	0.195	0.383	0.556	0.707
x	5π/16	3π/8	7π/16	π/2	
y	0.831	0.924	0.981	1	

$A = \tfrac{1}{2}\cdot\tfrac{\pi}{16}\{0 + 1 + 2(0.195 + 0.383 + 0.556 + 0.707$
$+ 0.831 + 0.924 + 0.981)\}$
$= 0.996\,866619 \approx 0.996\,87$ (5 d.p.)

In your table write your y-values correct to 3 d.p. but for greatest accuracy remember to use exact values in your calculations.

DON'T FORGET
When calculating the y-values make sure your calculator is in radian mode.

EXAM QUESTION 2

●●●

The diagram shows part of the curve $y = Ax^3 + Bx^2 + Cx + D$. (a) Find the values A, B, C, D. (b) Use the trapezium rule (5 ordinates) to estimate the value of the shaded area to 5 d.p. (c) Use algebraic integration to find the exact area of the shaded region. (d) Calculate the percentage error in using the trapezium rule to estimate this area, and say whether this estimate is an under- or overestimate.

LINKS
See the factor theorem on p. 25.

Curve crosses the x-axis at x = −1, x = 1 and x = 3, so (x+1), (x−1) and (x−3) must be factors
∴ y = k(x+1)(x−1)(x−3) = k(x²−1)(x−3)
= k(x³ − 3x² − x + 3)

When x = 0, y = 3 (from graph) so k = 1
∴ y = x³ − 3x² − x + 3
∴ A = 1, B = −3, C = −1, D = 3

x-values go from 3 to 4 so with 5 ordinates h = 0.25

x	3	3.25	3.5	3.75	4
y	0	2.391	5.625	9.797	15

Area = $\tfrac{1}{2}(0.25)\{0 + 15 + 2(2.391 + 5.625 + 9.797)\}$
$= 6.328125 \approx 6.328\,13$ unit² (5 d.p.)

$\int_3^4 (x^3 - 3x^2 - x + 3)\,dx = [\tfrac{1}{4}x^4 - x^3 - \tfrac{1}{2}x^2 + 3x]_3^4$
$= (\tfrac{1}{4}\cdot4^4 - 4^3 - \tfrac{1}{2}\cdot4^2 + 3\cdot4) - (\tfrac{1}{4}\cdot3^4 - 3^3 - \tfrac{1}{2}\cdot3^2 + 3\cdot3)$
$= 4 - (-\tfrac{9}{4}) = 4 + 2\tfrac{1}{4} = 6\tfrac{1}{4}$ unit²

% error = 100 × (6.328125 − 6.25)/6.25 = 1.25%
It is an overestimate

© Pearson Education Limited 2001

Integration – volumes of revolution

(AS) MEI-P1 CCEA-A2 AQA(A) EDEXCEL OCR-P2 WJEC-P3

When the area under a curve is rotated 360° about an axis, the volume of the solid of revolution can be found by integration.

THE FORMULAE

If the curve is rotated about the x-axis then $V = \int \pi y^2 dx$

If the curve is rotated about the y-axis then $V = \int \pi x^2 dy$

Fill in the two formulae you need to know.

ROTATION ABOUT THE X-AXIS

Find the volume of the solid formed when the area bounded by the curve $y = 2x^2 + 3x$ is rotated through 360° about the x-axis between the ordinates $x = 1$ and $x = 2$; leave your answer in terms of π.

Step 1 If possible, sketch the curve to see where it cuts the axes.

when $y = 0$, $0 = 2x^2 + 3x$
$0 = x(2x + 3)$
$x = 0$ or $-\frac{3}{2}$

Complete step 1.

Step 2 Formula uses y^2 so find y^2 in terms of x. Have $y = 2x^2 + 3x$ so

$y^2 = (2x^2 + 3x)^2 = 4x^4 + 12x^3 + 9x^2$

DON'T FORGET
To find y^2, put $2x^2 + 3x$ in a bracket then square the bracket.

Complete step 2.

Step 3 Now substitute y^2 into $V = \int \pi y^2 dx$ and integrate.
$V = \pi \int_1^2 (4x^4 + 12x^3 + 9x^2) dx = \pi[\frac{4}{5}x^5 + 3x^4 + 3x^3]_1^2$

EXAMINER'S SECRETS
Try to work in fractions and leave π in your answer; it means your answer will be accurate and it looks good too.

Complete step 3.

Step 4 Substitute the limits and evaluate.
$V = \pi(\frac{4}{5}.2^5 + 3.2^4 + 3.2^3) - \pi(\frac{4}{5}.1^5 + 3.1^4 + 3.1^3)$
$= \pi[\frac{128}{5} + 48 + 24 - (\frac{4}{5} + 3 + 3)] = 90\frac{4}{5}\pi$ unit^3

Complete step 4.

ROTATION ABOUT THE Y-AXIS

Find the volume of the solid formed when the area bounded by the curve $y = x^3 + 2$ (in the first quadrant) and the line $y = 3$ is rotated through 360° about the y-axis.

Step 1 Draw a sketch of the curve if possible.
Sketching this curve is best done by translating the curve $y = x^2$ through 2 units, i.e. move the curve up 2 units

Complete step 1.

Step 2 Formula uses x^2 so find x^2 in terms of y. Have $y = x^2 + 2$ so
$x^2 = y - 2$

Complete step 2.

Step 3 Now substitute x^2 into $V = \int \pi x^2 dy$ and integrate.
$V = \pi \int_2^3 (y - 2) dy = \pi[\frac{1}{2}y^2 - 2y]_2^3$

Complete step 3.

Step 4 Substitute the limits and evaluate.
$V = \pi(\frac{1}{2}.3^2 - 2.3) - \pi(\frac{1}{2}.2^2 - 2.2) = \frac{1}{2}\pi$ unit^3

Complete step 4.

Turn the page for some exam questions on this topic ▶

For more on this topic, see pages 60–61 of the *Revision Express A-level Study Guide*

EXAM QUESTION 1

Find the volume of the solid formed when the area enclosed by the curve $y = x - 1/x$, the x-axis and the line $x = 2$ is rotated through 360° about the x-axis.

For volume we need $\int \pi y^2 dx$
Now $y = x - 1/x$ so $y^2 = (x - 1/x)^2$
$\therefore y^2 = x^2 - 2x(1/x) + (1/x)(1/x) = x^2 - 2 + x^{-2}$
$\therefore$ Vol $= \pi \int_1^2 (x^2 - 2 + x^{-2}) dx = \pi[\frac{1}{3}x^3 - 2x - x^{-1}]_1^2$
$= \pi(\frac{1}{3}.2^3 - 2.2 - \frac{1}{2}) - \pi(\frac{1}{3}.1^3 - 2.1 - \frac{1}{1})$
$= \pi(\frac{7}{3} - \frac{3}{2}) = \frac{5}{6}\pi$ unit^3

DON'T FORGET
If the question asks for the answer correct to 2 d.p. or 3 s.f. then evaluate your final answer, otherwise leave π as it is.

EXAM QUESTION 2

(a) Sketch the curve $y = 2x^2 + 1$ and mark any points where the curve crosses the axes. (b) Find the area enclosed by the curve, the y-axis, the x-axis and the line $x = 3$. (c) Find the volume of the solid formed when the area in the first quadrant bounded by the curve, the y-axis and the line $y = 3$ is rotated through 360° about the y-axis.

EXAMINER'S SECRETS
Even if you're not asked for a sketch, it's often a good idea to draw something. A sketch can help you see how to approach the question and choose the limits for the integration.

(a) Sketch the curve by stretching $y = x^2$ by scale factor 2, then translating through 1 unit, i.e. move the curve up 1 unit

(b) Area $= \int y\, dx$
$= \int_0^3 (2x^2 + 1) dx$
$= [\frac{2}{3}x^3 + x]_0^3$
$= (\frac{2}{3}.3^3 + 3) - (\frac{2}{3}.0^3 + 0)$
$= 18 + 3 = 21$ unit^2

Calculate the area.

(c) Volume $= \int \pi x^2 dy$
Now $y = 2x^2 + 1 \therefore x^2 = \frac{1}{2}(y - 1)$
$\therefore$ Volume $= \frac{1}{2}\pi \int_1^3 (y - 1) dy$
$= \frac{1}{2}\pi[\frac{1}{2}y^2 - y]_1^3$
$= \frac{1}{2}\pi(\frac{1}{2}.3^2 - 3) - \frac{1}{2}\pi(\frac{1}{2}.1^2 - 1)$
$= \frac{1}{2}\pi(\frac{3}{2} + \frac{1}{2}) = \pi$ unit^3

Now calculate the volume.

Proof

AQA(A) AQA(B) EDEXCEL MEI OCR WJEC-P1

When trying to prove something, make sure your working is clear, and check each step is logical and correct.

NOTATION USED

Match each word statement with its corresponding symbol statement.

1	P is equal to Q	$P = Q$
2	P is implied by Q	$P \neq Q$
3	P is approximately equal to Q	$\neg Q$
4	Not Q	$P \equiv Q$
5	P implies Q	$P \not\Rightarrow Q$
6	P is not equal to Q	$P \Rightarrow Q$
7	P does not imply Q	$P \Leftarrow Q$
8	P is identical to Q	$P \approx Q$

○○○ ③ ⑥ ④ ⑧ ⑦ ⑤ ② ①

DIRECT PROOF

See if you can finish the proof.

Direct proof starts with facts which are accepted, then argues logically to the result which is required.

Prove that if a is odd and b is even then ab is even. Start by defining a and b in terms of another letter to do with a being odd and b being even. If a is odd it can be written as $a = 2p + 1$. If b is even it can be written as $b = 2q$. Substituting into ab we have
$(2p + 1) \times 2q = 4pq + 2q = 2(2pq + q)$
$2(2pq + q)$ is divisible by 2 so ab must be even

○○○

PROOF BY COUNTEREXAMPLE

Find a counterexample for each of these.

DON'T FORGET
You just need to find one example which doesn't work.

To prove a statement is false, all we have to do is produce just one case which doesn't fit the statement.

$x^2 = y^2 \Rightarrow x = y$
$x = 1, y = -1$
For any two real numbers a, b we have $a/b = b/a$
$a = 2, b = 1$
For any two real numbers a, b we have $a - b > 0 \Rightarrow a^2 - b^2 > 0$
$a = 1, b = -2$

○○○

PROOF BY CONTRADICTION

SYLLABUS CHECK
Proof by counterexample and contradiction are required on the following modules: MEI-P1, AQA(A)-P2, Edexcel-P2, OCR-P3, WJEC-P2, AQA(B)-P4.

Now try to prove that the negated statement is false.

Start by negating the statement you want to prove then show this negation is false; this then means the statement must be true.
Prove by contradiction that if $f(x) = x^2 + bx + c$ then $b^2 - 4c < 0 \Rightarrow f(x) > 0$ for all real values of x.
Start by negating the statement, i.e. $b^2 - 4c < 0 \Rightarrow f(x) \leq 0$ for at least one real value of x.

Now $b^2 - 4c < 0 \Rightarrow f(x) = 0$ has non-real roots and $f(x) \leq 0$
for at least one real value of x
$\Rightarrow f(x) = 0$ has at least one real value of x
$\Rightarrow f(x)$ has real roots
Therefore we have a contradiction

○○○

Turn the page for some exam questions on this topic ▶

For more on this topic, see pages 34–35 of the *Revision Express A-level Study Guide*

EXAM QUESTION 1

Prove that if a and b are both odd numbers, then $a + b$ is even.

Start by defining a and b in terms of other letters.

If a is odd it can be written in the form $a = 2m + 1$
If b is odd it can be written in the form $b = 2n + 1$
Therefore $a + b = 2m + 1 + 2n + 1$
$= 2m + 2n + 2 = 2(m + n + 1)$
which is a multiple of 2, so $(a + b)$ is even

EXAM QUESTION 2

Prove the identity $(\sin x + \cos x)^2 + (\sin x - \cos x)^2 = 2$.

Start with one side then prove it's equal to the other.

Decide which side to start with
Here the left-hand side is better
LHS $= (\sin x + \cos x)^2 + (\sin x - \cos x)^2$

$= \sin^2 x + 2\sin x \cos x + \cos^2 x + \sin^2 x - 2\sin x \cos x$
$\quad + \cos^2 x$

$= 2\sin^2 x + 2\cos^2 x$

$= 2(\sin^2 x + \cos^2 x) \quad (\cos^2 x + \sin^2 x = 1)$

$= 2 \times 1$

$= 2$

$=$ RHS

LINKS
See trigonometric equations on p. 21.

EXAM QUESTION 3

Prove by contradiction that $x + 1/x \geq 2$ for all $x > 0$.

Start by negating the statement
Assume $x + 1/x < 2$ for all $x > 0$
Rearrange the inequality to obtain $x^2 + 1 < 2x$
$x^2 - 2x + 1 < 0$
Find the critical values by solving
$x^2 - 2x + 1 = 0$
$(x - 1)(x - 1) = 0 \quad \therefore x = 1$
A sketch of the graph shows that for values of x such that
$x^2 - 2x + 1 < 0$ there is no solution as the graph doesn't go below the x-axis
For all $x > 0$ our assumption was not true; we've reached a contradiction
So $x + 1/x \geq 2$ for all $x > 0$

LINKS
See inequalities on p. 11.

Turn the page for some exam questions on this topic ▶

Algebra – surds, indices and laws of logs

You need to know how to use indices, surds and logarithms.

INDICES

Draw a line to connect the expression on the left with its corresponding expression on the right.

1. $a^m a^n$ — ③ a^{mn}
2. a^m/a^n — ⑥ $\sqrt[n]{a}$
3. $(a^m)^n$ — ④ 1
4. a^0 — ⑤ $1/a^n$
5. a^{-n} — ① a^{m-n}
6. $a^{1/n}$ — ② a^{m-n}

You need to rewrite 8 as a power of 2.

Using the laws of indices and without a calculator, find the exact value of $2^5 \times 8^3$.

$$2^5 \times 8^3 = 2^5 \times (2^3)^3 = 2^5 \times 2^{3\times3} = 2^5 \times 2^9$$
$$= 2^{5+9} = 2^{14}$$

SURDS

Surds are used to write down irrational numbers exactly.
You need to be able to simplify surds.
Simplify $\sqrt{32}$ and $(2+\sqrt{2})(3-\sqrt{8})$

THE JARGON
When asked to 'simplify' a surd, you need to write it with the smallest possible integer under the square root sign. You are looking to rewrite the surd in the form $\sqrt{x^2 y}$ which can be simplified to $x\sqrt{y}$.

$$\sqrt{32} = \sqrt{16 \times 2} = \sqrt{16} \times \sqrt{2} = 4\sqrt{2}$$
$$(2+\sqrt{2})(3-\sqrt{8}) = 2\times3 - 2\times\sqrt{8} + 3\times\sqrt{2} - \sqrt{2}\times\sqrt{8}$$
$$= 6 - 2\sqrt{2\times4} + 3\sqrt{2} - \sqrt{2\times8}$$
$$= 6 - 4\sqrt{2} + 3\sqrt{2} - \sqrt{16}$$
$$= 6 - \sqrt{2} - 4 = 2 - \sqrt{2}$$

THE JARGON
Rationalizing the denominator means eliminating any surds in the denominator.

You need to be able to rationalize the denominator.
Simplify $\frac{\sqrt{3}}{\sqrt{2}}$ by multiplying by $\frac{\sqrt{2}}{\sqrt{2}}$.

$$\frac{\sqrt{3}}{\sqrt{2}} = \frac{\sqrt{3}}{\sqrt{2}}\frac{\sqrt{2}}{\sqrt{2}} = \frac{\sqrt{6}}{\sqrt{4}} = \frac{\sqrt{6}}{2}$$

LAWS OF LOGARITHMS

Write down the three laws of logarithms.

The power rule
$$\log x^n = n \log x$$
The multiplication rule
$$\log(pq) = \log p + \log q$$
The division rule
$$\log\left(\frac{p}{q}\right) = \log p - \log q$$

For more on this topic, see pages 4–5 and 25 of the *Revision Express A-level Study Guide*.

EXAM QUESTION 1

Express $\frac{8}{\sqrt{8}} - \frac{2}{\sqrt{2}}$ in the form $a\sqrt{b}$ where a and b are integers.

Take your time and don't rush. Rationalize the denominator for each expression then simplify the fractions and the surds to obtain the answer.

$$\frac{8}{\sqrt{8}} - \frac{2}{\sqrt{2}} = \frac{8}{\sqrt{8}}\frac{\sqrt{8}}{\sqrt{8}} - \frac{2}{\sqrt{2}}\frac{\sqrt{2}}{\sqrt{2}} = \frac{8\sqrt{8}}{8} - \frac{2\sqrt{2}}{2}$$
$$= \sqrt{8} - \sqrt{2} = \sqrt{4\times2} - \sqrt{2}$$
$$= 2\sqrt{2} - \sqrt{2} = \sqrt{2}$$
so $a = 1$ and $b = 2$

EXAM QUESTION 2

Solve the equation $2^x = 10^{5x-1}$; give your answer to 3 s.f.

Take logs of both sides then use the power rule to rewrite the equation with the unknowns in front of the logarithms. Now you can solve the equation in the normal way. You'll need to use your calculator to evaluate any logarithms in your answer.

Take logs of both sides
$$\log 2^x = \log 10^{5x-1}$$
$$x\log 2 = (5x-1)$$
$$5x - x\log 2 = 1$$
$$x(5 - \log 2) = 1$$
$$x = \frac{1}{5 - \log 2} = 0.213 \ (3\ \text{s.f.})$$

EXAM QUESTION 3

The strength of a particular radioactive source after t years is given by $R = 8000 \times 5^{-0.003t}$. State the initial value of R and find the value of t when the source has decayed to half its value; give your answer correct to three significant figures.

The initial value is found when $t = 0$.

Initial value is when $t = 0$
so $R = 8000 \times 5^0 = 8000$

Write an equation with R equal to half the initial value.

$$8000 \times 5^{-0.003t} = 4000$$
$$5^{-0.003t} = 0.5$$

Take logs of both sides then solve.

Take logs of both sides
$$\log 5^{-0.003t} = \log 0.5$$
$$-0.003t\log 5 = \log 0.5$$
$$-0.003t = \frac{\log 0.5}{\log 5}$$
$$-0.003t = -0.430\ 676\ 56$$
$$t = 143.558\ 85$$
$$t = 144 \ \text{years} \ (3\ \text{s.f.})$$

Algebra – e^x and $\ln x$

You need to be familiar with the exponential function and the related logarithmic function.

THE EXPONENTIAL FUNCTION e^x

The exponential function has a very important property.

$$\frac{d}{dx}e^x = e^x$$

> Write down the important property of e^x.

GRAPHING e^x AND $\ln x$

$$y = e^x$$

$$y = \ln x$$

> Sketch the graphs of e^x and $\ln x$ on the axes provided.

EXAMINER'S SECRETS
You are sketching a function and its inverse. This can be done by reflecting the function in the line $y = x$.

LINKS
See laws of logarithms on p. 39.

THE NATURAL LOGARITHM $\ln x$

The inverse function of e^x is $\ln x$; $\ln x$ obeys all the laws of logarithms.

Find x when $e^{3x} = 7$.

Take natural logs of both sides

$$\ln e^{3x} = \ln 7$$

$$3x = \ln 7$$

$$x = \frac{\ln 7}{3}$$

$$x = 0.649 \ (3 \ d.p.)$$

> You'll need to use logarithms for this.

DIFFERENTIATING e^x AND $\ln x$

Differentiating e^x is very easy. But what about differentiating $\ln x$?

$$\frac{d}{dx}e^{4x} = 4e^{4x}$$

$$\frac{d}{dx}(\ln x) = \frac{1}{x}$$

> Differentiate these functions.

Turn the page for some exam questions on this topic ▶

For more on this topic, see pages 24–25 of the *Revision Express A-level Study Guide*

EXAM QUESTION 1

In a lab experiment the relationship between the number of bacteria N present in a culture after time t hours is modelled by $N = 50e^{1.3t}$.

(a) When will the number of bacteria have reached 10 000? (b) What will be the rate of increase of bacteria per hour when $t = 8$? Give your answers to two significant figures.

(a) First find the time when $N = 10\,000$

$$N = 10\,000$$

$$10\,000 = 50e^{1.3t}$$

$$200 = e^{1.3t}$$

Take natural logs of both sides

$$\ln 200 = 1.3t$$

$$t = \frac{\ln 200}{1.3}$$

$$t = 4.1\,h$$

(b) Now find the rate of increase dN/dt when $t = 8$

$$N = 50e^{1.3t}$$

$$\frac{dN}{dt} = 1.3 \times 50e^{1.3t}$$

Substitute $t = 8$

$$\frac{dN}{dt} = 65e^{1.3 \times 8} = 2.1 \times 10^6$$

EXAM QUESTION 2

The equation of a curve is $y = 4x^2 - 2\ln x$ where $x > 0$. Find the coordinates of the stationary point on the curve.

Stationary point when $dy/dx = 0$

$$y = 4x^2 - 2\ln x$$

$$\frac{dy}{dx} = 8x - \frac{2}{x}$$

$$\frac{dy}{dx} = 0 \Rightarrow 8x - \frac{2}{x} = 0 \Rightarrow 8x = \frac{2}{x}$$

$$x^2 = \frac{1}{4}$$

$$x = 0.5 \text{ or } x = -0.5$$

$$as \ x > 0, \ x = 0.5$$

$$y = 4(0.5)^2 - 2\ln(0.5) = 2.39 \ (3 \ s.f.)$$

The stationary point is therefore $(0.5, 2.39)$.

DON'T FORGET
Stationary points or turning points occur when the gradient is zero.

LINKS
See differentiation on p. 29.

© Pearson Education Limited 2001

Sequences and series

AS AQA(A) AQA(B) CCEA EDEXCEL OCR(A) MEI WJEC

You are studying lists of numbers constructed from a formula or an inductive definition.

SEQUENCE OR SERIES

Define sequence and series.

Sequence
A list of numbers (terms) in a defined order with a rule for generating them.

Series
The terms of a sequence added together.

EXAMINER'S SECRETS
Familiarize yourself with the different notation for sequences and series.

GENERATING SEQUENCES AND SERIES

You need to be able to generate a sequence from a formula or an inductive definition, and a series from its sigma notation.

Formula definition
Write the first five terms defined by formula $u_n = 2n^2 - 1$.
1, 7, 17, 31, 49

Inductive definition
Write the first five terms generated by $u_1 = 3$, $u_{n+1} = 3u_n + 2$.
3, 11, 35, 107, 323

Sigma definition
Write out the series $\sum_{r=1}^{n} 2r^3$ up to its fifth term.
2 + 16 + 54 + 128 + 250

THE JARGON
Sigma is the Greek letter Σ. It means 'the sum of'.

ARITHMETIC AND GEOMETRIC PROGRESSIONS

Circle the correct formula.
An arithmetic progression (AP) has first term a and common difference d. The nth term is found using which formula?
$(n - 1)d$ $a + nd$ $a + (n + 1)d$ $\boxed{a + (n - 1)d}$

Circle the correct formula.
The sum of the first n terms S_n can be found using which formula?
$n(2a + d)$ $\tfrac{1}{2}n(2a + nd)$ $\boxed{\tfrac{1}{2}n[2a + (n - 1)d]}$ $\tfrac{1}{2}n[2a + (n + 1)d]$

Circle the correct formula.
A geometric progression (GP) has first term a and common ratio r. The nth term is found by using which formula?
$\boxed{ar^{n-1}}$ ar^n $a^{n-1}r$ $a + (n - 1)r$

Circle the correct formula.
The sum S_n of the first n terms can be found using which formula?
$a(1 - r^n)$ $\boxed{\dfrac{a(1 - r^n)}{1 - r}}$ $\dfrac{a(1 - n)}{1 - r}$ $a + (n - 1)r$

THE JARGON
Convergence is when a sequence or series tends to a limiting value.

Write down the condition for convergence.
Some geometric progressions converge and the sum to infinity is found by using the formula $a/(1 - n)$. This can only be used when r satisfies a certain condition.
$-1 < r < 1$

What is the sum of the numbers 1 to 100?
Arithmetic progression with $a = 1$, $d = 1$
$\therefore S_n = 0.5(100)(2 + 99)$
$= 5050$

Turn the page for some exam questions on this topic ▶

For more on this topic, see pages 28–31 of the *Revision Express A-level Study Guide*

EXAM QUESTION 1

The fourth term of an arithmetic progression is five times the first term, and the second term is 7. Find the first term, the common difference and the sum of the first ten terms.

To answer this question, use the formula for the *n*th term of an AP hence set up two simultaneous equations in a and d.

$u_4 = 5u_1$ $u_2 = 7$
$a + 3d = 5a$ $a + d = 7$
$4a = 3d$ (1) $a = 7 - d$ (2)
Substitute (2) into (1)
$4(7 - d) = 3d$
$28 - 4d = 3d$
$7d = 28$
$d = 4$ so $a = 3$
$S_{10} = 0.5(10)[2(3) + (10 - 1)4]$
$= 210$

EXAM QUESTION 2

A mortgage is taken out for £20 000 and is repaid by annual instalments of £4000. Interest is charged on the outstanding debt at 10% calculated annually.

If the first repayment is made one year after the mortgage is taken out, find the number of years it takes to repay the mortgage.

Write expressions for the amount outstanding after 1, 2 and 3 years.

Let m_n be the amount outstanding after n years
$m_1 = 20\,000(1.1) - 4000$
$m_2 = [20\,000(1.1) - 4000](1.1) - 4000$
$m_2 = 20\,000(1.1)^2 - 4000(1.1) - 4000$
$m_3 = [20\,000(1.1)^2 - 4000(1.1) - 4000](1.1) - 4000$
$m_3 = 20\,000(1.1)^3 - 4000(1.1)^2 - 4000(1.1) - 4000$

Now write down the amount outstanding after n years.

$m_n = 20\,000(1.1)^n - 4000(1.1)^{n-1}$
$\quad - 4000(1.1)^{n-2} - \ldots - 4000(1.1) - 4000$
$= 20\,000(1.1)^n - 4000[1 + (1.1) + (1.1)^2$
$\quad + \ldots + (1.1)^{n-1} + (1.1)^{n-1}]$

Simplify your expression and look for a GP. Rewrite your GP using the formula for the first *n* terms.

The term in square brackets is the first n terms of a geometric progression where $a = 1$ and $r = 1.1$; $m_n = 0$ when the loan is repaid.

After *n* years you want the outstanding amount to be zero. Use this to write an equation you can solve by logarithms.

$20\,000(1.1)^n - 4000 \dfrac{(1.1)^n - 1}{1.1 - 1} = 0$

$20\,000(1.1)^n = 4000 \dfrac{(1.1)^n - 1}{1.1 - 1}$

$5(0.1)(1.1)^n = (1.1)^n - 1$ so $(1.1)^n = 2$

$\log(1.1)^n = \log 2$ so $n = \dfrac{\log 2}{\log 1.1} = 7.3$

Mortgage is repaid in 8 years (n must be integer)

LINKS
See logarithms on p. 39.

Functions

(AS) AQA(A) AQA(B) CCEA EDEXCEL OCR(A) MEI WJEC

The idea behind functions such as $y = x^2$, $y = x^3 - 2$ and $y = 2x$ is that for every value of x, you can find a unique value of y.

DOMAIN AND RANGE

Define domain and range.

Domain
The input of the function

Range
The output of the function

Find the range of function $y = \sqrt{x(10 - x)}$ with domain $0 \le x \le 10$.

What does the root sign tell you about y?

$y \ge 0$ by definition of square root

Square both sides; form a quadratic in x.

Squaring both sides
$y^2 = x(10 - x) \Rightarrow y^2 = 10x - x^2$
$x^2 - 10x + y^2 = 0$

By considering the discriminant write down an inequality for y and hence write down the range of the function.

For real roots the discriminant is greater than or equal to zero
$(-10)^2 - 4(1)y^2 \ge 0 \Rightarrow y^2 \le 25 \Rightarrow y \le 5$
The range is $0 \le y \le 5$

LINKS
See quadratics on p. 5.

EXAMINER'S SECRETS
To find an inverse function first write the function as $y = f(x)$. Now make x the subject of the formula. Then swap x and y. You have found the inverse function.

INVERSE FUNCTIONS

Find the inverse of the function $f(x) = \dfrac{2x + 3}{x}$, $x \ne 0$.

Let $y = \dfrac{2x + 3}{x}$

Make x the subject of the formula

$xy = 2x + 3$

$xy - 2x = 3$

$x(y - 2) = 3$

$x = \dfrac{3}{y - 2} \quad \therefore f^{-1}(x) = \dfrac{3}{x - 2}$

TRANSFORMATIONS OF GRAPHS

If a curve is defined by the function $y = f(x)$, we can transform the graph of y by changing the expression involving $f(x)$.

Draw a line connecting the new f(x) expression on the left to the resulting transformation on the right.

1	$f(x - a)$	Stretch in the y-direction, scale factor a
2	$f(x) + a$	Translation of a units in the x-direction
3	$f(ax)$	Translation of a units in the y-direction
4	$af(x)$	Stretch in the x-direction, scale factor $1/a$

④ ① ② ③

Turn the page for some exam questions on this topic ▶

For more on this topic, see pages 16–21 of the *Revision Express A-level Study Guide*

EXAM QUESTION 1

Sketch the function $y = x^2$. Hence sketch graphs of (a) $y = (x + 2)^2$ and (b) $y = 4x^2 + 1$. Show clearly where the graphs cross the axes.

EXAMINER'S SECRETS
A sketch is just that. If it does not have to be an accurate plot but it must show the essential features. Label the axes and the graph and show the points of intersection between the curve and the axes. Label any other points which are useful in giving more information about the graph.

First sketch the function $y = x^2$

$y = x^2$

Think of each graph as a transformation of the original function then sketch it.

(a) Now sketch $y = (x + 2)^2$
This is a translation
of -2 units in the
x-direction

$y = (x + 2)^2$

(b) And finally $y = 4x^2 + 1$
$y = 4x^2 + 1$
$y = (2x)^2 + 1$
This is a stretch in the
x-direction, scale factor
0.5 followed by a
translation of 1 unit in
the y-direction

$y = 4x^2 + 1$
$(1, 5)$

EXAM QUESTION 2

For what value of x is the function $f(x) = 3/(2 - x)$ undefined?

EXAMINER'S SECRETS
In an exam be prepared for questions on function ranges and domains.

Find the inverse function $f^{-1}(x)$.

$f(x)$ is undefined when $x = 2$

Let $y = \dfrac{3}{2 - x}$

Now make x the subject of the formula

$(2 - x) = \dfrac{3}{y} \Rightarrow x = 2 - \dfrac{3}{y}$

$f^{-1}(x) = 2 - \dfrac{3}{x}$

Differentiation – further functions

For more on this topic, see pages 52–53 of the *Revision Express A-level Study Guide*

(AS) AQA(A) AQA(B) CCEA EDEXCEL OCR(A) MEI WJEC

Having mastered the basics, you need to apply them in finding equations of tangents and normals, and you need to know how to differentiate products and quotients.

PRODUCT RULE AND QUOTIENT RULE

These rules are easy to remember in terms of u and v, where u and v are both functions of x.

The product rule tells you how to differentiate the product $y = uv$.

$$\frac{dy}{dx} = v\frac{du}{dx} + u\frac{dv}{dx}$$

The quotient rule tells you how to differentiate the quotient $y = u/v$.

$$\frac{dy}{dx} = \frac{1}{v^2}\left(v\frac{du}{dx} - u\frac{dv}{dx}\right)$$

Differentiate $y = e^x(4x^2 + 3)$ with respect to x.

$$\frac{dy}{dx} = e^x(8x) + (4x^2 + 3)(e^x) = e^x(4x^2 + 8x + 3)$$
$$= e^x(2x + 1)(2x + 3)$$

Differentiate $y = (2x + 1)/(3x - 1)$ with respect to x.

$$\frac{dy}{dx} = \frac{1}{(3x-1)^2}[(3x-1)(2) - (2x+1)(3)]$$
$$= \frac{-5}{(3x-1)^2}$$

Write the product rule here.

Write the quotient rule here.

LINKS
See e^x on p. 41.

CHAIN RULE

The chain rule is very useful for differentiating composite functions.

Find dy/dx when $y = (2 + x^3)^5$ by using the substitution $u = 2 + x^3$.

Chain rule: $\dfrac{dy}{dx} = \dfrac{dy}{du} \times \dfrac{du}{dx}$

Make the substitution $y = u^5$

$$\frac{dy}{du} = 5u^4 \qquad \frac{du}{dx} = 3x^2$$
$$\frac{dy}{dx} = 5u^4 \times 3x^2 = 15x^2(2 + x^3)^4$$

Begin by writing down the chain rule.

Substitute $u = 2 + x^3$ into the expression for y then apply the chain rule.

Turn the page for some exam questions on this topic ▶

EXAM QUESTION 1

Find the equation of the tangent to the curve $y = e^x(10 + 2x)$ when $x = -1$. Give your answers to 3 s.f.

First differentiate by using the product rule

$$\frac{dy}{dx} = e^x(2) + (10 + 2x)e^x = e^x(12 + 2x)$$

when $x = -1$, $\dfrac{dy}{dx} = e^{-1}[12 + 2(-1)] = 3.68$ (3 s.f.)

and $\quad y = e^{-1}[10 + 2(-1)] = 2.94$ (3 s.f.)

General equation of tangent is given by

$$y - y_1 = m(x - x_1) \text{ so } y - 2.94 = 3.68[x - (-1)]$$
$$y - 2.94 = 3.68(x + 1)$$
$$y = 3.68x + 6.62$$

which is the equation of the tangent at $x = -1$.

To find the equation of a tangent you first have to find the gradient. For this you will have to differentiate by using the product rule. You will also have to find the value of y when $x = -1$. From this information you can now find the equation of the tangent.

EXAM QUESTION 2

Find the coordinates of the stationary point on the curve $y = (3 - x)/x^2$ where $x > 0$, and determine whether it is a maximum or a minimum.

Using the quotient rule

$$\frac{dy}{dx} = \frac{x^2(-1) - (3-x)(2x)}{x^4} = \frac{x^2 - 6x}{x^4}$$

Turning points when $\dfrac{dy}{dx} = 0$

$$\Rightarrow x^2 - 6x = 0 \Rightarrow x(x - 6) = 0$$

$x = 0$ or $x = 6$

The curve is not defined when $x = 0$

so $x = 6$ is the only turning point

x	1	6	7
dy/dx	-5	0	0.003

Find y when $x = 6$
$y = (3 - 6)/6^2 = -\frac{1}{12}$
There is a minimum at the point $(6, -\frac{1}{12})$.

WATCH OUT
This curve has an asymptote.

LINKS
See differentiation on p. 29.

You need to find the y-coordinate.

Differentiation – higher derivatives

AS AQA(A) AQA(B) CCEA EDEXCEL OCR(A) MEI WJEC

This section deals with higher derivatives and the differentiation of trig functions.

HIGHER DERIVATIVES

When you continually differentiate a function you get higher derivatives of that function. If $f(x) = x^4 + 2x^3 - 4x^2 + 2x + 6$ find $f'(x)$, $f''(x)$ and $f'''(x)$.

$$f'(x) = 4x^3 + 6x^2 - 8x + 2$$
$$f''(x) = 12x^2 + 12x - 8$$
$$f'''(x) = 24x + 12$$

Write down $f'(x)$, $f''(x)$ and $f'''(x)$.

THE JARGON
$f'(x)$ is the first derivative of the function $f(x)$, $f''(x)$ is the second derivative, and so on. You can write dy/dx for $f'(x)$ and d^2y/dx^2 for $f''(x)$.

MAXIMUM AND MINIMUM

You can use the second derivative to find out the nature of a turning point on a curve.

$\dfrac{d^2y}{dx^2} < 0 \Rightarrow$ minimum, **maximum**, point of inflexion

$\dfrac{d^2y}{dx^2} > 0 \Rightarrow$ **minimum**, maximum, point of inflexion

Highlight the correct term.

DIFFERENTIATING SIN X AND COS X

Differentiating sin x and cos x is easy if you remember the rules.

$$\frac{d}{dx}(\sin x) = \cos x$$
$$\frac{d}{dx}(\cos x) = -\sin x$$

Complete these equations.

If $f(x) = \sin 2x$ find $f'(x)$, $f''(x)$ and $f'''(x)$.

Let $y = \sin 2x$ and $u = 2x$ so $y = \sin u$

$$\frac{du}{dx} = 2$$
$$\frac{dy}{du} = \cos u$$

Chain rule: $\dfrac{dy}{dx} = \dfrac{dy}{du} \times \dfrac{du}{dx}$

$$\frac{dy}{dx} = f'(x) = 2 \cos u = 2 \cos 2x$$

Using the same method gives

$$f''(x) = -2(2)\sin 2x$$
$$= -4 \sin 2x$$
$$f'''(x) = -2(4)\cos 2x$$
$$= -8 \cos 2x$$

The function sin 2x is a composite function. Either differentiate it directly or use the substitution $u = 2x$ and apply the chain rule.

LINKS
See the chain rule on p. 47.

Turn the page for some exam questions on this topic ▶

For more on this topic, see pages 52–53 of the *Revision Express A-level Study Guide*

EXAM QUESTION 1

If $y = 2x^3 - 3x^2 - 12x + 4$ find d^2y/dx^2, hence find the coordinates of the stationary points on the curve represented by this equation. Determine the nature of these stationary points.

$$\frac{dy}{dx} = 6x^2 - 6x - 12$$
$$\frac{d^2y}{dx^2} = 12x - 6$$

For turning points $\dfrac{dy}{dx} = 0$

$$6x^2 - 6x - 12 = 0$$
$$6(x^2 - x - 2) = 0$$
$$6(x - 2)(x + 1) = 0$$
$$x = 2 \text{ or } x = -1$$

When $x = 2$
$$y = 2(2)^3 - 3(2)^2 - 12(2) + 4 = -16$$
$$\frac{d^2y}{dx^2} = 12(2) - 6 = 18 \text{ so } \frac{d^2y}{dx^2} > 0$$

When $x = -1$
$$y = 2(-1)^3 - 3(-1)^2 - 12(-1) + 4 = 11$$
$$\frac{d^2y}{dx^2} = 12(-1) - 6 = -18 \text{ so } \frac{d^2y}{dx^2} < 0$$

$(2, -16)$ is a minimum and $(-1, 11)$ is a maximum

DON'T FORGET
You can determine the nature of turning points by another method, but use this question to practise the second derivative.

You also need to find the y-coordinate of each stationary point.

EXAM QUESTION 2

If $y = \tan x$ show that $dy/dx = \sec^2 x$.

$$y = \tan x = \frac{\sin x}{\cos x}$$

Using the quotient rule

$$\frac{dy}{dx} = \frac{\cos x(\cos x) - (\sin x)(-\sin x)}{\cos^2 x}$$
$$= \frac{\cos^2 x + \sin^2 x}{\cos^2 x}$$

Using $\cos^2 x + \sin^2 x = 1$
$$\frac{dy}{dx} = \frac{1}{\cos^2 x} = \sec^2 x$$

First write tan x in terms of sin x and cos x. Now use the quotient rule to find dy/dx. You'll need to use a trig identity to finish.

LINKS
See trigonometric equations on p. 21.

© Pearson Education Limited 2001

AS AQA(A) AQA(B) EDEXCEL OCR(A) MEI WJEC

Integration – further functions

Having mastered the basics, try integrating more complicated functions and finding volumes of revolution.

Substitution

Many functions can be integrated by using a substitution.

Find the integral of $(3x+1)^6$ by using the substitution $u = 3x+1$.

Complete step 1. Don't forget $\frac{dx}{du} = \frac{1}{du/dx}$

Step 1
Find du/dx and hence dx/du.

$\frac{du}{dx} = 3 \Rightarrow \frac{dx}{du} = \frac{1}{3}$

Complete step 2.

Step 2
Make the substitution and replace dx with $(dx/du)du$.

$\int (3x+1)^6 dx = \int u^6 \frac{dx}{du} du = \int u^6 \left(\frac{1}{3}\right) du = \frac{1}{3}\int u^6 du$

Complete step 3. Don't forget the constant of integration.

Step 3
Now everything should be in terms of u and du, so it can be integrated in the normal way.

$\frac{1}{3}\int u^6 du = \frac{1}{3}\left(\frac{u^7}{7}\right) + c = \frac{u^7}{21} + c$

Complete step 4. If you have a definite integral and limits are given, you need to change the limits in terms of u and leave out step 4.

Step 4
Finally substitute for u in terms of x.

$\int (3x+1)^6 dx = \frac{1}{21}(3x+1)^7 + c$

Volume of revolution

The volume of revolution of a curve about an axis can be found from these integrals:

Write down the two integrals.

For rotation about the x-axis $\int \pi y^2 dx$

For rotation about the y-axis $\int \pi x^2 dy$

Find the volume generated when the area enclosed by $y = x^2$, $x = 0$, $x = 2$ and $y = 0$ is rotated through 360° about $y = 0$.

$V = \int_0^2 \pi y^2 dx = \int_0^2 \pi x^4 dx \qquad (y = x^2 \text{ so } y^2 = x^4)$

$= \pi[\tfrac{1}{5}x^5]_0^2 = \tfrac{32}{5}\pi$

EXAMINER'S SECRETS
It's often helpful to leave π in your answer.

Turn the page for some exam questions on this topic ➤

For more on this topic, see pages 58–61 of the *Revision Express A-level Study Guide*

EXAMINER'S SECRETS
The word 'hence' means that you need to use whatever you have just done in order to answer the next part of the question.

LINKS
See the area under a curve on p. 31.

Exam question 1

Using the substitution $u = x^3 + 2$, show how $\int_0^1 x^2(x^3+2)^2 dx$ can be rewritten as $\frac{1}{3}\int_2^3 u^2 du$. Hence find the volume generated when the area enclosed by $y = x^2(x^3+2)$, $x = 0$, $x = 1$ and $y = 0$ is rotated through 360° about $y = 0$.

$u = x^3 + 2$

$\frac{du}{dx} = 3x^2 \qquad \frac{dx}{du} = \frac{1}{3x^2}$

change limits

x	u
0	2
1	3

$\int_0^1 x^2(x^3+2)^2 dx = \int_{x=0}^{x=1} x^2u^2 \frac{dx}{du} du$

$= \int_{x=0}^{x=1} x^2u^2 \frac{1}{3x^2} du = \frac{1}{3}\int_2^3 u^2 du$

$V = \int_0^1 \pi y^2 dx$ and $y^2 = [x(x^3+2)]^2 = x^2(x^3+2)^2$

so $V = \int_0^1 \pi x^2(x^3+2)^2 dx = \frac{1}{3}\pi\int_2^3 u^2 du$ (from above)

$= \frac{1}{3}\pi[\frac{1}{3}u^3]_2^3$

$= \frac{1}{9}\pi(3^3 - 2^3) = \frac{19}{9}\pi$

Exam question 2

Find the area bounded by $y = 1\sqrt{2x+1}$, $x = 4$, $x = 12$ and $y = 0$.

$A = \int_4^{12} (2x+1)^{-1/2} dx$

Let $u = 2x+1$ then $\frac{du}{dx} = 2$, $\frac{dx}{du} = \frac{1}{2}$

change limits

x	u
12	25
4	9

$A = \int_9^{25} u^{-1/2}\left(\frac{1}{2}\right) du$

$= \frac{1}{2}[2u^{1/2}]_9^{25} = (\sqrt{25} - \sqrt{9}) = 2$

© Pearson Education Limited 2001

Numerical methods – change of sign

AS AQA(A) AQA(B) CCEA EDEXCEL OCR(A) MEI WJEC

When you cannot solve an equation exactly you have to rely on numerical methods to find an approximate solution.

CHANGE OF SIGN

Locate where a root of a function lies by looking for a sign change.
Show that the function $f(x) = x^3 - 3x^2 + 1$ has a root between $x = 0$ and $x = 1$ when $f(x) = 0$.

$f(0) = 1$ $f(1) = 1 - 3 + 1 = -1$

There is a sign change between $f(0)$ and $f(1)$ therefore a root lies between $x = 0$ and $x = 1$

> Look for a sign change.

THE JARGON
A root of a function $f(x)$ is a value of x such that $f(x) = 0$.

DECIMAL SEARCH

Once you've located where a root lies, you can find an approximation for the root by using a decimal search.

For the function $f(x) = x^3 - 3x^2 + 1$, find to 1 d.p. the root of $f(x) = 0$ which lies in the interval $x = 0$ and $x = 1$.

We have $f(0) = 1$ and $f(1) = -1$ so choose $x_1 = 0.5$

> When looking for a root in the interval $x = a$ to $x = b$, all you need to do is show that $f(a)$ has a different sign to $f(b)$.

$f(x_1)$ =   0.375   too big/**too small**
x_2 =   0.6
$f(x_2)$ =   0.136   too big/**too small**
x_3 =   0.7
$f(x_3)$ =   –0.127   **too big**/too small
x_4 =   0.65
$f(x_1)$ =   0.007125   too big/**too small**

So the root lies between 0.65 and 0.7
Root = 0.7 to 1 d.p.

> Complete the stages then highlight 'too big' or 'too small', whichever is correct. At each iteration increase x by 0.1 until you find the solution.

> At each stage you need to decide whether your approximation is too big or too small. If the value is > 0 then the value is too big. If the value is too big and the interval changes sign from + to – then the value is too small.

NEWTON–RAPHSON

Newton–Raphson is far more efficient than interval bisection.

Given that $f(x) = x^3 - x^2 + 20x - 5$ has a root between $x = 0$ and $x = 1$, perform one iteration of the Newton–Raphson method to find the root to 1 d.p. using 0.5 as your first approximation.

Begin with $f'(x) = 3x^2 - 2x + 20$

x_1 = 0.5
$f(0.5)$ = 4.875
$f'(0.5)$ = 19.75

x_2 = 0.5 – (4.875/19.75)   = 0.5 – 0.2468
= 0.253 = 0.3 (1 d.p.)

> **DON'T FORGET**
> $x_{n+1} = x_n - f(x_n)/f'(x_n)$

> Complete the working.

LINKS
See differentiation on p. 29.

Turn the page for some exam questions on this topic ▶

EXAM QUESTION

The sketch graph shows $f(x) = x^3 - 8x^2 - 4x - 5$. It crosses the x-axis at s. The consecutive integers a and b are either side of s.

Find the values of a and b. Use the Newton–Raphson method to find the value of s to four significant figures.

First find a and b by looking for a sign change

$f(1) = -16$
$f(2) = -37$
$f(3) = -62$
...
$f(7) = -82$
$f(8) = -37$
$f(9) = +40$ (sign change)

Therefore $a = 8, b = 9$

Now use the Newton–Raphson method

$f'(x) = 3x^2 - 16x - 4$

$$x_{n+1} = x_n - \frac{x_n^3 - 8x_n^2 - 4x_n - 5}{3x_n^2 - 16x_n - 4}$$

Let $x_1 = 8.5$ (as it's halfway between 8 and 9)

$$x_2 = 8.5 - \frac{(8.5)^3 - 8(8.5)^2 - 4(8.5) - 5}{3(8.5)^2 - 16(8.5) - 4}$$
$$= 8.5375$$

$$x_3 = 8.5375 - \frac{(8.5375)^3 - 8(8.5375)^2 - 4(8.5375) - 5}{3(8.5375)^2 - 16(8.5375) - 4}$$
$$= 8.5371$$

$$x_4 = 8.5371 - \frac{(8.5371)^3 - 8(8.5371)^2 - 4(8.5371) - 5}{3(8.5371)^2 - 16(8.5371) - 4}$$
$$= 8.5371$$

Therefore $s = 8.537$ (4 s.f.)

> When answering this question you will need to look for a sign change. From the graph can you guess a possible value for s? Once you have found a and b, the midpoint is a sensible first approximation for Newton–Raphson.

WATCH OUT
Work to a sensible level of accuracy. The question asks you for four significant figures so work to at least five.

For more on this topic, see pages 64–67 of the Revision Express A-level Study Guide

For more on this topic, see pages 66–67 of the *Revision Express A-level Study Guide*

EXAM QUESTION 1

Show that the iterative formula
$$x_{n+1} = \sqrt[3]{\frac{15 - x_n}{x_n^2}}$$

can find a root of the equation $x^4 + x - 15 = 0$ and that the equation has a root between $x = 1$ and $x = 2$. Start with $x_1 = 2$ then use the formula to find $x_2, x_3, \ldots, x_6$ giving your answers to five decimal places. Say whether this sequence is divergent or convergent.

Iterative formula

$$x_{n+1} = \sqrt[3]{\frac{15 - x_n}{x_n^2}} \Rightarrow x = \sqrt[3]{\frac{15 - x}{x^2}}$$

$$\Rightarrow x^3 = \frac{15 - x}{x^2} \Rightarrow x^5 = 15 - x$$

$$\Rightarrow x^5 + x - 15 = 0$$

Let $f(x) = x^5 + x - 15$
$f(1) = -13$ $f(2) = 19$
A root must lie between $x = 1$ and $x = 2$ as there is a sign change

$x_1 = 2$

$x_2 = \sqrt[3]{\dfrac{15 - (2)}{2^2}} = 1.48125$

$x_3 = \sqrt[3]{\dfrac{15 - (1.48125)}{1.48125^2}} = 1.83327$

$x_4 = \sqrt[3]{\dfrac{15 - (1.83327)}{1.83327^2}} = 1.57643$

$x_5 = \sqrt[3]{\dfrac{15 - (1.57643)}{1.57643^2}} = 1.75458$

$x_6 = \sqrt[3]{\dfrac{15 - (1.75458)}{1.75458^2}} = 1.62645$

The sequence is converging but very slowly

You will need to rearrange the iterative formula until you get the original equation. Look for a sign change to see that the interval contains a root.

LINKS
See change of sign on p. 53.

Now you can investigate to see whether the sequence is diverging or converging.

EXAM QUESTION 2

Use rearrangement to decide which of these iterative formulae could be tried when finding a root of the equation $x^4 - 2x^2 + x - 1 = 0$.

(a) $x_{n+1} = \sqrt{\dfrac{1 - x_n}{x_n^2 - 2}}$ (b) $x_{n+1} = \sqrt{\dfrac{x_n}{x_n^2 + 2}}$ (c) $x_{n+1} = \dfrac{1}{x_n^3 - 2x_n + 1}$

It is possible to use (a) and (c)

Numerical methods – convergence

There are other ways of finding numerical solutions to equations. Not all numerical methods work. Sometimes a method will not give a solution but will lead to divergence.

REARRANGING IN THE FORM $x = g(x)$

This method often finds a root of an equation $f(x) = 0$ by rewriting it as $x = g(x)$.

The root α of the equation $x^3 - 4x - 7 = 0$ can be found by using the iterative formula
$$x_{n+1} = \sqrt[3]{4x_n + 7}$$

Taking $x_1 = 2.5$ find x_8 and write down α correct to 4 s.f.

$x_1 = 2.5$
$x_2 = \sqrt[3]{4(2.5) + 7} = 2.57128$
$x_3 = \sqrt[3]{4(2.57128) + 7} = 2.58558$
$x_4 = \sqrt[3]{4(2.58558) + 7} = 2.58843$
$x_5 = \sqrt[3]{4(2.58843) + 7} = 2.58899$
$x_6 = \sqrt[3]{4(2.58899) + 7} = 2.58910$
$x_7 = \sqrt[3]{4(2.58910) + 7} = 2.58913$
$x_8 = \sqrt[3]{4(2.58913) + 7} = 2.58913$

$\alpha = 2.589$ (4 s.f.)

Try out this formula using your calculator.

To show that an iterative formula finds the root of a given equation, write it without the subscripts n and $n + 1$. Now rearrange until you get the equation you're asked to solve.

STAIRCASE AND COBWEB DIAGRAMS

Not all rearrangements work, so be able to recognize convergence and divergence. They can be represented graphically.

Convergence   Convergence   Divergence   Divergence

Below each diagram say whether it shows divergence or convergence to the root α.

Turn the page for some exam questions on this topic ▶

Vectors

AQA(A) AQA(B) CCEA EDEXCEL OCR(A) MEI

Vectors have magnitude and direction; scalars, have only magnitude. For example, a speed of 5 m s⁻¹ is a scalar (it only has magnitude) but a velocity of 5 m s⁻¹ vertically downwards is a vector (it has magnitude and direction).

REPRESENTING VECTORS
○○○

You can use a line with an arrow to represent a vector. The length of the line represents the magnitude and the arrow shows the direction of the vector.

> Draw lines to represent the following vectors: (a) a velocity v magnitude 3 m s⁻¹ and direction north-east, (b) a displacement AB of magnitude 4 m and direction 30° below the horizontal.

RESULTANT VECTORS
○○○

A resultant vector is the sum of two or more vectors. You can find the resultant by using a vector triangle.

Two forces of 4 N and 2 N have an angle of 40° between them. Find to 2 d.p. the magnitude of the resultant vector.

> The angle between two vectors is the angle made when both vectors are drawn facing away from each other and starting from the same point. To make a vector triangle, redraw the vectors 'nose to tail' so that one vector starts where the other vector finishes. The resultant vector is the third side of the triangle, and the angle between the vectors is the supplementary angle. Find the magnitude of the resultant vector by using the cosine rule.

$r^2 = 4^2 + 2^2 - 2 \times 4 \times 2 \cos 140°$
$= 32.2567 \therefore r = 5.68\,N$

Find to 2 d.p the magnitude and direction of the vector 6i − 2j.

> If the vector is described by components i and j, use Pythagoras for its magnitude and trigonometry for its direction angle.

$r = \sqrt{6^2 + 2^2} = 6.32$

$\theta = \tan^{-1}(-2/6) = -18.43°$

RESOLVING VECTORS
○○○

You need to be able to resolve a vector into two perpendicular components. F is a force of 4 N at 30° to the horizontal. Resolve F into horizontal and vertical components.

> To resolve a vector, draw a right-angled triangle with the vector as the hypotenuse. Line up the other sides of the triangle with the directions in which you're resolving; use trigonometry to find the component sides of the triangle.

Horizontally $F = 4 \cos 30°$
$= 3.46\,N$

Vertically $F = 4 \sin 30°$
$= 2\,N$

Turn the page for some exam questions on this topic ▶

For more on this topic, see pages 158–159 of the *Revision Express A-level Study Guide*

EXAM QUESTION 1

The diagram shows the forces acting on a body. Resolve the forces parallel to OX and then parallel to OY. Hence find the resultant force on the body and state its direction, giving your answer to 2 d.p.

> Find the total force in the x-direction and the total force in the y-direction.

∥ to OX
$6 + 3 \cos 30° - 2 \cos 60° - 7 \cos 45° = 2.648$

∥ to OY
$4 + 3 \sin 30° + 2 \sin 60° - 7 \sin 45° = 2.282$

> Now find the resultant using Pythagoras and the direction using trigonometry. A sketch will help at this stage.

$r^2 = 2.648^2 + 2.282^2$
$r = \sqrt{12.219} = 3.50$
$\theta = \tan^{-1}\left(\frac{2.282}{2.648}\right) = 40.75°$

Resultant force is 3.50 N at an angle of 40.75° above OX

EXAM QUESTION 2

Two boys are pulling a sled using two ropes attached to the same point on the sled. The ropes are parallel with the ground and make an angle of 50° with each other. If the tensions in the rope are 5 N and 3 N, what is the magnitude of the resultant pulling force?

> Draw a sketch and a vector triangle.

$r^2 = 5^2 + 3^2 - 2 \times 5 \times 3 \cos 130°$
$= 25 + 9 - 30 \cos 130°$
$= 53.284$
$r = 7.30\,N$
Resultant force is 7.3 N

Kinematics

Kinematics is the study of motion. You are considering a body moving in a straight line with a constant acceleration a. The body starts with an initial velocity u and finishes with a final velocity v, covering a displacement s in a time t.

GRAPHS

○○○

The motion of a body can be described by a displacement–time graph or a velocity–time graph.

1 Displacement–time graph   Gradient = velocity   **1**

2 Velocity–time graph   Gradient = acceleration   **2**

    Area = displacement   **2**

> **Connect each type of graph on the left with its corresponding statements on the right.**

Sarah throws a ball vertically up into the air and catches it again. Which of these graphs could be the velocity–time graph for the ball?

A   (B)   C   D

> **Circle the correct letter.**

DON'T FORGET
Acceleration is constant here.

EQUATIONS OF MOTION

The four equations of motion relate the variables u, v, s, t and a.

$v = u + at$ $s = ut + \frac{1}{2}at^2$

$v^2 = u^2 + 2as$ $s = \frac{1}{2}(u + v)t$

> **Write down the four equations of motion.**

DON'T FORGET
Each of the equations leaves out one of the variables.

VERTICAL MOTION UNDER GRAVITY

○○○

You are applying the equations of motion vertically in a straight line. The only acceleration is the constant acceleration due to gravity g.

$v = u + at$

$v = 0$ at the ball's greatest height

> **Write down the equation of motion needed to find the time taken for a ball to reach its greatest height. What can you say about the value of v at this height?**

Turn the page for some exam questions on this topic ➤

For more on this topic, see pages 160–161 of the *Revision Express A-level Study Guide*

EXAM QUESTION 1

A particle is moving with a velocity of $8\,\mathrm{m\,s^{-1}}$ in a straight line. At a time $t = 0$ it is subjected to an acceleration of $-4\,\mathrm{m\,s^{-2}}$ for 4 s and then continues at a constant velocity for 3 s before being brought to rest by a constant deceleration in 6 s. Draw a velocity–time graph then find **(a)** the total distance covered and **(b)** the total increase in displacement.

DON'T FORGET
Distance is not the same as displacement. Displacement is a vector, so it can be positive or negative; distance is a scalar.

> **Your graph should use straight lines only. Displacement on a velocity–time graph is the area under the curve.**

(a) Find the total distance covered

$Total\ distance = triangle's\ area + trapezium's\ area$

$$= \frac{8 \times 2}{2} + \frac{8(3 + 11)}{2} = 64\,\mathrm{m}$$

(b) Find the increase in displacement

$Displacement = triangle's\ area - trapezium's\ area$

$$= 8 - 56 = -48\,\mathrm{m}$$

EXAM QUESTION 2

○○○

A particle is projected upwards with a speed of $16\,\mathrm{m\,s^{-1}}$ from point A. At the same time a second particle is released from rest 8 m directly above A. Find d, the distance above A, and the time taken for the particles to collide (take $g = 9.8\,\mathrm{m\,s^{-2}}$).

> **If d is the distance above A where the particles collide, then for the second particle $s = 8 - d$.**

Particle at A   Particle 8 m above A

$u = 16, a = -9.8$ $u = 0, a = 9.8, s = 8 - d$

$s = ut + \frac{1}{2}at^2$ $s = ut + \frac{1}{2}at^2$

$d = 16t - 4.9t^2$ (1) $8 - d = 0 + 4.9t^2$

$d = 8 - 4.9t^2$ (2)

> **Now solve the simultaneous equations.**

Substitute (2) into (1)

$8 - 4.9t^2 = 16t - 4.9t^2$

$8 = 16t$

$t = 0.5\,s$

Substitute $t = 0.5\,s$ into (2)

$d = 8 - 4.9(0.5)^2$

$= 6.775\,\mathrm{m}$

Force

AS AQA(A) AQA(B) CCEA EDEXCEL OCR(A) MEI WJEC

Forces are very important in mechanics. A force is necessary to make an object begin to move or to bring a moving object to rest. Forces have magnitude and direction and are therefore vectors. The unit of force is the newton (N).

DIFFERENT TYPES OF FORCE

○○○

Name these common forces.

Force due to gravity
Weight

Forces found in strings and springs
Tension and thrust

Contact forces
Friction and normal reaction

FORCE DIAGRAMS

○○○

Many mechanics questions can be answered using a force diagram.

A block of weight W rests on a smooth slope inclined at 20° to the horizontal. The block is held at rest by a rope parallel to the slope. Which diagram correctly shows all the forces acting on the block?

Circle the correct letter.

A   B   C   D

Draw a force diagram to represent a book resting on a table. The book is subjected to a horizontal force P and the book is at rest.

Use filled arrows to represent the forces and label them with appropriate letters, e.g. W for weight and F for friction.

Draw a force diagram to represent a person standing in a lift that is accelerating upwards.

Turn the page for some exam questions on this topic ➤

For more on this topic, see pages 168–169 of the *Revision Express A-level Study Guide*

EXAM QUESTION 1

● ● ●

The diagram shows a rough plank resting on a block with one end of the plank on rough ground.

Draw diagrams to show

(a) the forces acting on the plank
(b) the forces acting on the block

These questions would only be the start of an exam question. You would probably be asked to find the magnitude of some of the forces when given some more information.

(a) Forces acting on the plank

(b) Forces acting on the block

EXAM QUESTION 2

● ● ●

Two bricks A and B have weights 15 N and 7 N respectively. B rests on a table and A rests on B. Draw diagrams to show the forces acting (a) on brick A and (b) on brick B.

(a) Forces on brick A

(b) Forces on brick B

Newton's laws of motion

(AS) AQA(A) AQA(B) CCEA EDEXCEL OCR(A) MEI WJEC

Isaac Newton's laws form the basis of mechanics. You need to be able to quote these laws and use them in solving problems.

NEWTON'S THREE LAWS ○○○

Newton's first law

Every body will remain at rest or continue to move with a uniform velocity in a straight line unless external forces act on it.

Newton's second law

The force acting on a body of constant mass is directly proportional to the acceleration produced.

Newton's third law

If a body A exerts a force on a body B, then B exerts an equal and opposite force on A.

Write down Newton's three laws in the spaces provided.

$F = ma$ ○○○

A car of mass 1400 kg is pushed by a force of 200 N. Calculate the acceleration of the car.

$$F = ma$$
$$a = \frac{F}{m} \quad \text{therefore } a = \frac{200}{1400} = \frac{1}{7}\,\text{m s}^{-2}$$

Here is a force diagram of a person of mass M kg standing in a lift. The lift is moving vertically upwards with a constant acceleration.

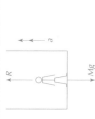

Which equation correctly describes the motion?

(A) $R = Mg$

(B) $Mg - R = Ma$

(C) $R - Mg = Ma$

(D) $Ma - R = Mg$

The correct equation is (C)

Most of your calculations will use Newton's second law. Use Newton's second law and rearrange to find a.

DON'T FORGET
Weight = mg

DON'T FORGET
You can use a double-arrowhead to represent acceleration.

Turn the page for some exam questions on this topic ▶

For more on this topic, see pages 168–171 of the *Revision Express A-level Study Guide*

EXAM QUESTION 1 •••

A car of mass 450 kg is brought to rest in a time of 5 s from a speed of 25 m s⁻¹.

Assuming the braking force is constant and assuming there is no resistance to motion, find the force exerted by the brakes.

$v = u + at$
$0 = 25 + 5a$
$a = -5\,\text{m s}^{-2}$

This is a deceleration.
It acts in the same direction as the force exerted by the brakes

$u = 25\,\text{m s}^{-1}$

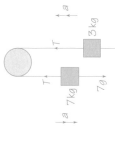

$a = -5\,\text{m s}^{-2}$

$F = ma$
$F = 450 \times 5$
$F = 2250\,\text{N}$

You'll need to find a by using a constant acceleration equation. A diagram will help you decide the directions for the braking force and the acceleration.

THE JARGON
The assumptions made in a model are the simplifications such as no air resistance or lack of friction.

LINKS
See kinematics on p. 59.

EXAM QUESTION 2 •••

Two particles of mass 3 kg and 7 kg are connected by a light inelastic string passing over a smooth fixed pulley. The system is released from rest.

Say how these two assumptions affect the model: (a) the pulley is smooth and (b) the string is inelastic.

A smooth pulley means tension either side of the pulley will be the same.
An inelastic string means the acceleration of the two particles will be the same.

Now find the acceleration of the particles and the tension in the string; leave g in your answer.

EXAMINER'S SECRETS
When dealing with connected particles, always consider each particle separately. A carefully labelled diagram is essential for answering this question.

First draw a diagram.

For 7 kg mass $7g - T = 7a$ (1)
For 3 kg mass $T - 3g = 3a$ (2)

$(1) + (2) \Rightarrow 4g = 10a$
$a = \frac{2}{5}g\,\text{m s}^{-2}$

Substitute into (2)
$T - 3g = 3 \times \frac{2}{5}g$
$T = (3 \times \frac{2}{5}g) + 3g = \frac{21}{5}g\,\text{N}$

Using F = ma write down two equations, one for each particle, then solve them simultaneously.

Projectiles

AS AQA(A) AQA(B) CCEA EDEXCEL OCR(A) MEI WJEC

Any body moving only under the action of its own weight after being given an initial velocity u is a projectile. We consider bodies projected at some given angle to the horizontal.

LINKS

See kinematics on p. 59.

EQUATIONS

Resolve the motion into the x-direction (horizontal) and the y-direction (vertical). In the x-direction things are quite simple as there is no acceleration. In the y-direction there is acceleration due to gravity.

Equations in the x-direction

$u_x = u\cos\theta$

$x = ut\cos\theta$

> What is the component of the initial velocity u in the x-direction? Hence write down an equation for x, the displacement in the x-direction.

Equations in the y-direction

$u_y = u\sin\theta$

$y = ut\sin\theta - \frac{1}{2}gt^2$

$v_y = u\sin\theta - gt$

> What is the component of the initial velocity u in the y-direction? Hence, using the equations of motion, write down two equations in the y-direction, one for the displacement y and one for the velocity v, after time t.

MAXIMUM HEIGHT AND RANGE

Many problems involve finding the greatest height and the range (horizontal distance) of the projectile.

A particle is projected with a velocity of $40\,\text{m s}^{-1}$ at an angle of 30° to the horizontal. What is the greatest height reached by the particle and what is the range? Take $g = 10\,\text{m s}^{-2}$.

Greatest height

$v = u\sin\theta - gt$ $y = ut\sin\theta - gt^2$

$0 = 40\sin30° - gt$ $= (40 \times 2)\sin30° - 5 \times 2^2$

$gt = 20 \therefore t = 2$ $= 40 - 20 = 20\,\text{m}$

So the greatest height is 20 m

> For the greatest height you only need to think in the y-direction and let $v = 0$.

> Find t and substitute into the displacement equation.

Range

Time $= 2t = 4\,\text{s}$

$x = ut\cos\theta$

$x = (40 \times 4)\cos30° = 138.6\,\text{m}$

So the range is 138.6 m

> For the range, the time in the air will be twice the time taken to reach the greatest height, something you've already found. Substitute this time into the equation for x.

DON'T FORGET

The maximum range of a projectile is found when $\theta = 45°$.

For more on this topic, see pages 166–167 of the *Revision Express A-level Study Guide*

EXAM QUESTION 1

A particle is projected from a point O, with initial speed $V\,\text{m s}^{-1}$ at an angle θ above the horizontal. After 8 s the vertical component of the velocity is $10\,\text{m s}^{-1}$ downwards and the horizontal component is $18\,\text{m s}^{-1}$. Find to 2 d.p. the values of V and θ (take $g = 9.8\,\text{m s}^{-2}$).

We need to find u_y, the initial velocity in the y-direction

From $v = u + at$

$-10 = u_y - 9.8(8)$

$u_y = 78.4 - 10$

$= 68.4$

> find the initial velocity in the y-direction by using $v = u + at$.

The horizontal component of the velocity, $u_x = 18$, doesn't change

> You already know the initial velocity in the x-direction because it stays constant.

By Pythagoras's theorem

$V^2 = 18^2 + 68.4^2 = 5002.56$

$V = \sqrt{5002.56} = 70.7\,\text{m s}^{-1}$

> Find V by using Pythagoras.

Find θ by using the horizontal and vertical components of the velocity

$\tan\theta = \dfrac{68.4}{18} \therefore \theta = \tan^{-1}\left(\dfrac{68.4}{18}\right) = 75.26°$

> You can find θ by using trigonometry.

EXAM QUESTION 2

A particle is projected with speed $20\,\text{m s}^{-1}$ at 30° to the horizontal. Find to 1 d.p. the velocity of the particle after 4 s ($g = 9.8\,\text{m s}^{-2}$)

x-direction   y-direction

$v_x = u\cos\theta$ $v_y = u\sin\theta - gt$

$= 20\cos30°$ $= 20\sin30° - 9.8 \times 4$

$= 17.32$ $= 10 - 39.2 = -29.2$

$v = \sqrt{v_x^2 + v_y^2} = \sqrt{17.32^2 + (-29.2)^2} = 34.0\,\text{m s}^{-1}$

$\alpha = \tan^{-1}\left(\dfrac{-29.2}{17.32}\right) = -59.3°$

Velocity is $34\,\text{m s}^{-1}$ at 59.3° below the horizontal

DON'T FORGET

The final velocity will be the resultant of the vertical and horizontal components. When you have found these components you must combine them using Pythagoras to obtain the final velocity. You must also state the direction in which the particle is travelling (use trigonometry).

© Pearson Education Limited 2001

Friction

AQA(A) AQA(B) CCEA EDEXCEL OCR(A) MEI WJEC

Friction is the 'sticky' force that occurs whenever one surface slides over another. In mechanics, when modelling with friction, we say that the contact is rough.

THE FRICTION FORMULA

Say how friction acts in relation to the way an object is trying to move.

Friction always acts in the opposite direction to the one in which the object is trying to move.

Write an inequality for friction relating the friction F, normal reaction N and the coefficient of friction μ.

$$F \leq \mu N$$

Explain why this is an inequality.

Friction will increase, opposing any motion until the limiting point is reached when the object is about to move. At this limiting point, friction is at a maximum.

LIMITING FRICTION

A book of weight 4N rests on a rough table. It is pushed by a horizontal force P and the coefficient of friction is 0.75. Find the value of the frictional force and describe the motion for $P = 1$, $P = 3$ and $P = 5$.

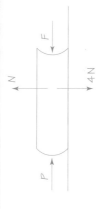

Resolving vertically $N = 4$

Resolving horizontally $F \leq \mu N$

     $F \leq 0.75 \times 4$

     $F \leq 3$

when $P = 1$, $F = 1$ so book doesn't move

when $P = 3$, $F = 3$ so book is about to move

when $P = 5$, $F = 3$ so book slides in direction of P

EXAMINER'S SECRETS
Only when friction is limiting will an object start to move.

EXAMINER'S SECRETS
Always start by drawing a good force diagram. You will need to find the normal contact force.

LINKS
See force on p. 61.

Turn the page for some exam questions on this topic ▶

For more on this topic, see pages 176–177 of the *Revision Express A-level Study Guide*

EXAM QUESTION 1

A toboggan of weight 80 N rests on a snow-covered slope inclined at an angle of 15° to the horizontal. Given that the coefficient of friction between the toboggan and the slope is 0.2, find whether the toboggan will slide. Give your answers to 2 d.p.

Resolving perpendicular to the slope

$N = 80 \cos 15° = 77.27$ N

Limiting value of friction is

$\mu N = 0.2 \times 77.27 = 15.45$ N

Weight component acting down the slope is

$80 \sin 15° = 20.71$ N

This weight component down the slope is greater than the maximum possible frictional force up the slope, so the toboggan will slide.

Draw a force diagram. Decide which force will make the toboggan slide and determine whether or not this force is greater than the limiting friction.

EXAM QUESTION 2

A suitcase of mass 40 kg is pulled using a rope inclined at 30° to the horizontal. If the coefficient of friction between the suitcase and the floor is 0.25, what is the least force needed to make the suitcase move? Take $g = 9.81$ m s⁻² and give your answers to 2 d.p.

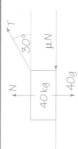

Friction is limiting $T \cos 30° = \mu N$

Resolving horizontally $T \cos 30° = 0.25 N$ (1)

Resolving vertically

$N + T \sin 30° = 40 g$

   $N = 40g - T \sin 30°$ (2)

Substitute (2) into (1)

$T \cos 30° = 0.25 \times 40g - 0.25 \times T \sin 30°$

$10g = T \cos 30° + 0.25 \times T \sin 30°$

$10g = T(\cos 30° + 0.25 \sin 30°)$

$T = \dfrac{10g}{\cos 30° + 0.25 \sin 30°} = 98.99$ N

EXAMINER'S SECRETS
Always determine whether or not the body is moving. If it is not moving, you must decide whether the friction is limiting.

Moments and equilibrium

When forces are in equilibrium the resultant is zero and there is no turning effect.

○○○

LAMI'S THEOREM

Finish this sentence.

Lami's theorem can only be used when *there are three forces acting in equilibrium.*

Draw a good force diagram and work out any missing angles. Write Lami's theorem then rearrange to find the tensions.

A weight of 20 N is hanging from two strings inclined at 30° and 60° to the vertical. What are the tensions in the strings? Answer to 2 d.p.

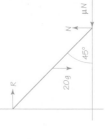

If there is more than one tension in a question, the chances are that they have different magnitudes. Give them different labels, e.g. T_1 and T_2.

Using the sine rule

$$\frac{20}{\sin 90°} = \frac{T_1}{\sin 120°} = \frac{T_2}{\sin 150°}$$

$T_1 = 20 \sin 120° = 17.32$ N

$T_2 = 20 \sin 150° = 10$ N

○○○

EQUILIBRIUM PROBLEMS

If more than three forces are present then resolving the forces and taking the resultant force as zero can solve equilibrium problems.

A particle of weight 5 N rests on a rough slope inclined at 30° to the horizontal. The coefficient of friction is 0.23 and the particle is prevented from sliding by an additional force H parallel to the slope. Given that the particle is on the point of sliding, find the value of H to 2 d.p.

First draw a force diagram then resolve parallel and perpendicular to the slope.

LINKS
See friction p. 67.

Resolving forces parallel to the slope

$\mu R + H - 5 \sin 30° = 0$ (1)

Resolving forces perpendicular to the slope

$R - 5 \cos 30° = 0$ ∴ $R = 4.33$ N

Substitute for R into (1)

$0.23 \times 4.33 + H - 2.5 = 0$

∴ $H = 1.50$ N

○○○

MOMENTS

Complete this definition.

The moment of a force about an axis is *the product of the force and the perpendicular distance from the axis.*

Turn the page for some exam questions on this topic ▶

For more on this topic, see pages 174–175 and 180–181 of the *Revision Express A-level Study Guide*

EXAM QUESTION 1

● ● ●

A uniform rod *AB* of length 2 m and weight 30 N rests horizontally on smooth supports at *A* and *B*. A weight of 5 N is attached to the rod at a distance of 0.3 m from *A*. Find to 2 d.p. the forces exerted on the rod by the supports.

Taking moments about appropriate axes can easily solve this problem. Choose an axis that eliminates one of the unknowns from your equation.

DON'T FORGET
The normal reactions at the two ends of the rod are not the same.

Taking moments about A

$(R_B \times 2) - (30 \times 1) - (5 \times 0.3) = 0$

$R_B = 15.75$ N

Taking moments about B

$(30 \times 1) + (5 \times 1.7) - (R_A \times 2) = 0$

$R_A = 19.25$ N

EXAM QUESTION 2

● ● ●

A ladder, which is 4 m long and has a mass of 20 kg, leans against a smooth vertical wall with its foot in contact with a rough horizontal floor. The ladder makes an angle of 45° with the horizontal and is on the point of slipping. Find the reaction between the wall and the ladder and find the coefficient of friction between the ladder and the floor. Take $g = 9.8\ \text{ms}^{-2}$.

Draw a good force diagram showing dimensions as well as forces.

Resolving vertically

$N = 20g = 20 \times 9.8 = 196$

Resolve forces into two perpendicular directions.

Resolving horizontally

$R = \mu N = 196\mu$ (1)

Remember that friction is limiting.

because friction is limiting

Take moments about a convenient axis. As the forces are in equilibrium the total moment is zero.

Taking moments about the foot of the ladder

$20g \times 2 \cos 45° - R \times 4 \sin 45° = 0$ (equilibrium)

$R = 98$

You should now be able to solve your equations.

Substitute for R in (1)

$196\mu = 98$

$\mu = 0.5$

Momentum and impulse

AS AQA(A) AQA(B) CCEA EDEXCEL OCR(A) MEI WJEC

When two balls collide or a bat hits a ball, what happens to the velocities and the forces during and after the impact?

DEFINITIONS

> **Give one definition of momentum and two definitions of impulse.**

Momentum
The product of a body's mass and velocity (mv)

Impulse: definition 1
The change in momentum $(mv - mu)$

Impulse: definition 2
The product of the force acting and the time for which it acts (Ft)

> **DON'T FORGET**
> The units of momentum and impulse are newton-seconds (Ns).

IMPULSE

With questions involving impulse you may need to carry out calculations with velocities or with force.

> **DON'T FORGET**
> All masses must be in kilograms. Convert 80 g to 0.08 kg.

A bat strikes a cricket ball of mass 80 grams giving it a velocity of 45 m s⁻¹. Find the impulse on the ball.
$$\text{Impulse} = mv - mu$$
$$= (0.08 \times 45) - (0.08 \times 0)$$
$$= 3.6 \text{ Ns}$$

> **DON'T FORGET**
> There are two forms of the impulse equation. You may need to use one or other, or both.

A particle of mass 3 kg is travelling with a velocity of 6 m s⁻¹. What constant force acting in the same direction of the particle will increase its speed to 30 m s⁻¹ in 4 s?
$$\text{Impulse} = mv - mu = (3 \times 30) - (3 \times 6) = 72$$
Then using impulse $= Ft$ gives us $F \times 4 = 72$
$$F = 18 \text{ N}$$

CONSERVATION OF MOMENTUM

> **Explain conservation of momentum.**

The total momentum after a collision is the same as the total momentum before the collision.

A ball of mass 8 kg is moving with a velocity of 5 m s⁻¹. It collides with a ball of mass 3 kg which is at rest. After the collision the 8 kg ball is moving at 2 m s⁻¹ in the same direction as before. Find the velocity of the smaller ball after the collision.

> **Draw a diagram showing the masses and velocities before and after the collision.**

Before: 5 → 8 kg, O, 3 kg
After: 2 → 8 kg, u → 3 kg

Momentum before $= 8 \times 5 = 40$
Momentum after $= (8 \times 2) + (3 \times u)$
By conservation of momentum
$(8 \times 2) + (3 \times u) = 40 \therefore u = 8 \text{ m s}^{-1}$

Turn the page for some exam questions on this topic ▶

For more on this topic, see pages 190–191 of the *Revision Express A-level Study Guide*

EXAM QUESTION 1

A bullet is fired into a box of sand mounted on a trolley. The bullet has a mass of 15 grams and the mass of the trolley and sand is 4 kg. After the bullet is fired, the box with the bullet embedded in it moves off with a speed of 1.6 m s⁻¹. What was the speed of the bullet just before it hit the sand?

> **The mass of the bullet and the mass of the sand need to be added together when calculating the momentum after impact.**

Let u be the bullet speed
Conservation of momentum
$$0.015u = (4 + 0.015) \times 1.6$$
$$u = 428 \text{ m s}^{-1} \text{ (3 s.f.)}$$

Before: 15 g □, 4 kg, → u, → O
After: □ 15 g, 4 kg, → 1.6 m s⁻¹

EXAM QUESTION 2

Two particles of the same mass are travelling towards each other along the same line with constant speeds 6 m s⁻¹ and 2 m s⁻¹. If they collide and coalesce, find their joint speed just after impact.

> **THE JARGON**
> Coalesce means the particles stick together and become one body.

> **If two bodies are approaching each other, you need to take one velocity as positive and the other as negative.**

Let m be the mass of the particles
Momentum
before $6m - 2m = 4m$
after $(m + m)u = 2mu$
Conservation of momentum
$$2mu = 4m \Rightarrow u = 2 \text{ m s}^{-1}$$

Before: (m) 6 m s⁻¹ → , (m) → 2 m s⁻¹
After: (2m) → u

EXAM QUESTION 3

Particles A, B and C lie in a straight line and have masses $4m$, $3m$ and m respectively. Initially B and C are at rest with A projected towards B with a speed of 5 m s⁻¹. After A collides with B, the speed of A is 2 m s⁻¹ in the same direction as before. When B collides with C the particles coalesce and move off with a joint speed v. Assuming no resistance to motion, find the value of v and the speed of B just before colliding with C.

> **Tackle the collisions one at a time using the conservation of momentum for each.**

Let B's speed after the first collision be u
First collision
$$4m \times 5 = (4m \times 2) + (3m \times u)$$
$$u = 4 \text{ m s}^{-1}$$
Second collision
$$3m \times 4 = (3m + m)v$$
$$v = 3 \text{ m s}^{-1}$$

Before Collision 1: 5 m s⁻¹ → (4m), O (3m)
After Collision 1: 2 m s⁻¹ → (4m), u → (3m)
Before Collision 2: 2m s⁻¹ → (3m), O (m)
After Collision 2: v → (4m)

Graph theory

AS AQA(A) AQA(B) EDEXCEL OCR MEI

Graphs in decision and discrete mathematics are not like the (x, y) grids normally associated with graphs. They are networks with vertices (or nodes) and edges (or arcs).

TYPICAL GRAPHS USED IN GRAPH THEORY ○○○

Describe each of these graphs in the spaces provided.

Planar graphs
A planar graph will have no edges crossing.

WATCH OUT
There are more graphs than are noted here. You may also be expected to show the information from a network in a table.

Simple graph
A simple graph has no multiple edges or loops.

Complete graphs
Complete graphs are simple and each pair of vertices is joined by an edge.

Bipartite graphs
A bipartite graph can have its vertices divided into two subsets. No edge connects vertices within the same subset.

Weighted graphs
In a weighted graph each edge will have a number associated to it.

LINKS
See spanning trees on p. 83.

Isomorphic graphs
Isomorphic graphs show the same connectedness.

Trees
Trees are connected graphs which have no cycles (or circuits).

EULER AND HIS TRAILS ○○○

Highlight Euler's formula

Euler's formula is

$v + e + f = 2$ $v - e + f = 2$

$f + e = v$ $v - e - f = 1$

REVISION EXPRESS
Finding out whether a graph is traversable by using degrees is on p. 87 of the *Revision Express A-level Study Guide.*

Traversability is where every edge of a network may be traced only once without taking pen from paper. There are two types of trail.

An eulerian trail will finish where it starts (it is closed).   True

State whether each sentence is true or false.

A semi-eulerian trail will not finish where it started (it is not closed).   True

Eulerian and semi-eulerian trails are not traversable.   False

K_N – A COMPLETE GRAPH WITH N VERTICES ○○○

Next to the diagram of K_5 show that K_4 is planar.

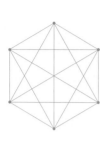

IF YOU HAVE TIME
Using Euler's formula, show that K_5 is non-planar.

Turn the page for some exam questions on this topic ▶

For more on this topic, see pages 86–87 of the *Revision Express A-level Study Guide*

EXAM QUESTION 1 ...

(a) Draw the graph of K_6. How many edges does it have?

(b) How many edges does the graph of K_n have?

First draw the graph

Try to work out the relationship between the number of vertices and how many they are connected to. Also, think of how many vertices are needed to make an edge.

Now work out the numbers of edges

K_6 has 15 edges

K_n has $\frac{1}{2}n(n-1)$ edges

EXAM QUESTION 2 ...

For each graph show whether it is (a) traversable with an eulerian trail, (b) traversable with a semi-eulerian trail or (c) not traversable.

By sketching over the graphs see if they are traversable. State whether they are eulerian trails, semi-eulerian trails or not traversable.

Not traversable

Eulerian trail

Eulerian trail

Semi-eulerian trail

DON'T FORGET
You can use the orders of the vertices to state whether or not a graph is traversable. Try a few and see what happens. You are looking to see how many edges meet at each vertex and whether the number is odd or even.

If all vertices are of even order then the graph will be traversable with an eulerian trail. If two vertices are of odd order and the rest are even then the graph is a semi-eulerian trail.

Spanning trees

AQA(A) AQA(B) EDEXCEL OCR MEI

Spanning trees are connected graphs which have no circuits. Problems associated with spanning trees may be about cabling towns, for example, and finding the minimum set-up costs. To solve these problems we need to find minimum spanning trees.

ALGORITHMS FOR FINDING MINIMUM SPANNING TREES

Prim's algorithm

Prim's algorithm is a greedy algorithm	True
Prim's algorithm starts with the shortest edge	False
Prim's algorithm chooses any vertex to be the starting point	True

Kruskal's algorithm

Kruskal's algorithm is a greedy algorithm	True
Kruskal's algorithm starts with the shortest edge	True
Kruskal's algorithm chooses any vertex to be the starting point	False

Prim's algorithm

Start with any vertex. Choose from this vertex the shortest edge (if there are two or more then choose randomly). Keep choosing the shortest edges from the vertices of the minimum spanning tree, ensuring circuits are not made, until there is one less edge than the number of vertices.

Kruskal's algorithm

Start with the shortest edge. Then choose the shortest edge remaining which does not make a circuit with the edges already chosen. (If there are two or more the same then choose at random.) Keep going until there is one less edge than the number of vertices.

NETWORKS IN TABLES

	A	B	C	D
A	–	7	8	5
B	7	–	2	4
C	8	2	–	4
D	5	4	4	–

Suppose you start at A. The shortest edge is AD, so delete columns A and D to prevent circuits from being formed. The shortest length from A or D is DC. Therefore delete column C. The last edge would therefore be from A, D or C and is CB, length 2.

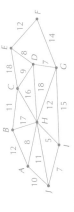

Minimum spanning tree

State whether each sentence is true or false.

State whether each sentence is true or false.

Set out the steps in Prim's algorithm.

Set out the steps in Kruskal's algorithm.

EXAMINER'S SECRETS
Learn both algorithms off by heart.

Computers can be used to find a minimum spanning tree, although the graph needs to be converted to a table first. Complete the table for this network.

Using these steps for Prim's algorithm, delete columns in the table with a highlighter until you're left with just one column. This process will produce the edges for a minimum spanning tree.

Draw the minimum spanning tree.

Turn the page for some exam questions on this topic ➤

For more on this topic, see pages 88–89 of the *Revision Express A-level Study Guide*

EXAM QUESTION 1

A model railway is set up with places for the train to visit as shown in the network. Using (a) Prim's algorithm and (b) Kruskal's algorithm work out how to connect all the places so that the total length of railway track is a minimum, starting from A.

[network diagram with vertices A, B, C, D, E, F, G, H, I, J and edge weights 12, 11, 18, 8, 9, 12, 16, 17, 18, 7, 14, 10, 8, 11, 5, 12, 15, 7]

(a) Prim's algorithm

Starting at A
$A \rightarrow H\ (8) \rightarrow I\ (5) \rightarrow J\ (7)$
Then $A \rightarrow B\ (12) \rightarrow C\ (11) \rightarrow D\ (9) \rightarrow G\ (7)$
Then $D \rightarrow E\ (8) \rightarrow F\ (12)$.
Minimum total length of track is 79 units

(b) Kruskal's algorithm

Starting with $I–H$ *(5 units, the shortest edge). Then* $D–G$ (7), $I–J$ (7), $A–H$ (8), $D–E$ (8), $D–C$ (9), $C–B$ (11), $E–F$ (12), $A–B$ (12)
Minimum total length of track is 79 units

Work out the minimum spanning tree using Prim's algorithm, and write down its total length.

EXAMINER'S SECRETS
Include enough information to show the examiner you've used Prim's algorithm. The final tree on its own isn't enough.

Work out the minimum spanning tree using Kruskal's algorithm, and write down its total length.

DON'T FORGET
A problem may have more than one minimum spanning tree, but the minimum length will always be the same.

EXAM QUESTION 2

This table shows a complete graph K_5. How can you tell that it's a complete graph?
There are values for every edge from any vertex to another.
The only vertex connections that aren't present are P to P, Q to Q, etc.

Draw a minimum spanning tree of the information.
The total length of the minimum spanning tree is 22 units.

	P	Q	R	S	T
P	–	8	12	13	7
Q	8	–	5	10	14
R	12	5	–	6	11
S	13	10	6	–	4
T	7	14	11	4	–

Copy out the table and, using Prim's algorithm, obtain a minimum spanning tree. Draw the minimum spanning tree and write down its total length.

Shortest paths

Shortest path problems do not require you to go along every edge nor do they expect you to visit every node. All they require is that you get from one specified node to another by the shortest route.

DIJKSTRA'S ALGORITHM TO FIND SHORTEST PATHS ○○○

Step 1 Give the starting *vertex* a value of *zero* and then box the *zero* at this vertex.

Step 2 All the vertices that are joined to the *latest* boxed vertex need to be *temporarily* labelled with the time from the *starting* vertex.

Step 3 From the *temporary* labels that are written on the graph, choose the one with the *least* value. Make this label *permanent* by putting a box around it.

Step 4 Keep doing steps *2* and *3* until the vertex you are aiming to get to has been *boxed*. If at any later point a *temporary* label is reduced, use the reduced value. If it is not reduced then the label does not change.

> Fill in the missing words to complete Dijkstra's algorithm.

> **EXAMINER'S SECRETS**
> When you're using the algorithm make sure you write down enough information so it's very clear that you are using Dijkstra's algorithm.

AN EXAMPLE USING DIJKSTRA'S ALGORITHM ○○○

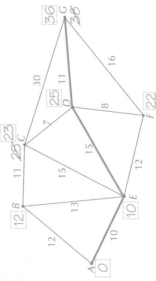

The shortest path is A to E to D to G.
The length of this path is 36 units.

> Following Dijkstra's algorithm write on this network to show how to label the vertices for finding the shortest path from A to G.

> Write down the shortest path and its length.

Turn the page for some exam questions on this topic ➤

For more on this topic, see pages 90–91 of the *Revision Express A-level Study Guide*

EXAM QUESTION

Here is a network of several towns and cities in France. The values are the distances between these places in kilometres. Work out the shortest path from Cherbourg to Marseille using Dijkstra's algorithm.

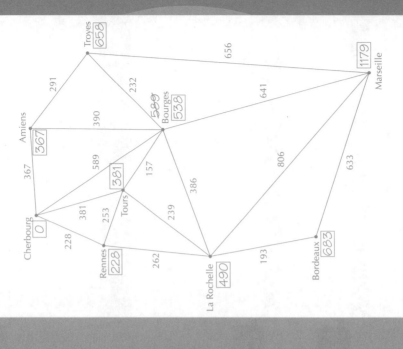

> **DON'T FORGET**
> Permanent labels are boxed; temporary labels are lightly crossed out.

> Write down the route for the shortest path and its length.

Shortest path is Cherbourg to Tours to Bourges to Marseille
Length = 1179 kilometres

Route inspection problem

A route inspection problem involves going along every edge and returning to the original vertex. The problem can easily be understood if you think of a postman, starting at the post office, delivering letters to all streets and then returning to the post office. Mei-Ko Kwan, a Chinese mathematician, was the first person to analyse this problem. That's why it's more commonly known as the Chinese postman problem.

CHINESE POSTMAN PROBLEM

Highlight the correct answers.

The route can be accomplished without repeating a street
if all the vertices are ~~even~~ /odd/mainly even.
New edges (or streets) ~~can be added~~ /cannot be added.
Vertices are made even by ~~adding a new edge~~/repeating an edge.

THE CHINESE POSTMAN ALGORITHM

Fill in the spaces to complete the steps of the Chinese postman algorithm.

Step 1 Ascertain which *vertices*, if any, are odd.

Step 2 Pair these *vertices* so that any extra distances added are kept to a minimum.

Step 3 Put in these repeated *edges* so that the graph formed has all its *vertices even*.

Step 4 Find the *eulerian* trail by inspection.

Apply the algorithm to this network; start and finish at A.

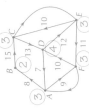

Write down the route and its length.

Vertices A, C, E, F are of odd order. By pairing A with F and C with E all vertices are of even order.

Route: $A \rightarrow B \rightarrow C \rightarrow D \rightarrow E \rightarrow C \rightarrow E \rightarrow F \rightarrow D \rightarrow A \rightarrow F \rightarrow A$.
Total length = 114 units

WATCH OUT
The route worked out is just one of the possible routes.

OTHER APPLICATIONS

Write down as many different applications as you can.

Route inspection algorithms have many other applications.

Printed circuit boards for electrical products
Road cleaning or inspection
Gritting roads during very cold weather
Running pipes, lines or cables underneath roads
Cleaning or checking sewers

Turn the page for some exam questions on this topic ▶

EXAM QUESTION

Here is a network of streets along which a newspaper delivery person has to deliver newspapers. They pick up the newspapers from A and they need to return there with any spares at the end of their delivery. Which route should be taken to keep it to a minimum? How long is this route? Distances are in metres.

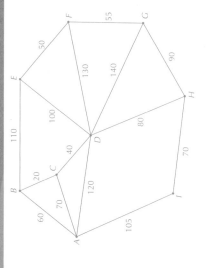

Draw out the network and add any extra edges required to complete the Chinese postman algorithm. Show the route.

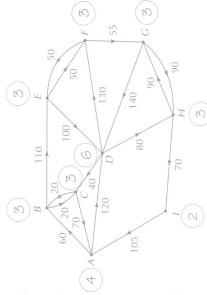

DON'T FORGET
Your route is just one possible route. You could try to find some other routes.

Write out the route.

Route
$A \rightarrow B \rightarrow C \rightarrow A \rightarrow D \rightarrow C \rightarrow B \rightarrow E \rightarrow D \rightarrow F \rightarrow E \rightarrow F \rightarrow$
$G \rightarrow D \rightarrow H \rightarrow G \rightarrow H \rightarrow I \rightarrow A$

Work out the length of the route.

Length
$60 + 20 + 70 + 120 + 40 + 20 + 110 + 100 + 130 + 50 +$
$50 + 55 + 140 + 80 + 90 + 90 + 70 + 105$
$= 1400$ metres

Travelling salesperson problem

AS AQA(A) AQA(B) EDEXCEL OCR MEI

The travelling salesperson problem (TSP) is similar to the Chinese postman problem except it involves visiting all the vertices rather than traversing all the edges. If the vertices are towns or cities and the edges are roads, rail tracks or air routes then it should be clear why it's known as TSP, with the traveller wanting to take the shortest route yet still visiting every vertex.

THE NEAREST-NEIGHBOUR ALGORITHM

Unfortunately, there is no simple algorithm that will automatically give you the shortest path. The nearest-neighbour algorithm will produce a route that is reasonable, but not necessarily the shortest.

You start at any vertex (usually stated) and then choose another vertex that is nearest to it and which has not yet been chosen. You keep choosing the nearest vertex until you have returned to the original vertex.

Explain the nearest-neighbour algorithm.

DON'T FORGET
There is more than one method for finding upper and lower bounds.

UPPER AND LOWER BOUNDS

The only way to be completely confident you have the optimal route for the TSP is to look at every route. Explain the reason for working out the upper and lower bounds.

Give a brief explanation why upper and lower bounds are calculated for the TSP.

They give a good idea of a range within which the optimal route must lie.

Let us obtain the upper bound and then the lower bound for this network.

Calculating the upper bound
Here is the minimum spanning tree for the network.

Draw the minimum spanning tree.

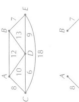

Traversing edges both ways will give an upper bound.

Redraw your minimum spanning tree with edges going both ways and write down the value of this upper bound.

$$2(8 + 6 + 9 + 7) = 60$$

Calculating the lower bound
After deleting vertex E (chosen randomly), we obtain the minimum spanning tree for the remaining vertices.

Draw the minimum spanning tree of the remaining vertices.

Adding the two shortest edges from E gives a lower bound.

Write down the values from the minimum spanning tree and add the two shortest edges from E

$$\underbrace{(12 + 8 + 6)}_{\text{minimum spanning tree}} + \underbrace{(9 + 7)}_{\text{edges from } E} = 42$$

Turn the page for some exam questions on this topic ▶

EXAM QUESTION 1

Fred is trying to sell his new lawnmower invention to businesses in different towns. Here is a network of the towns Fred wants to visit and the roads connecting them. Distances are in kilometres. Fred lives at A. Using the nearest-neighbour algorithm find a route that starts and finishes at A and visits every town; write down its length.

Show how to use the nearest-neighbour algorithm to work out a route.

EXAMINER'S SECRETS
Give enough detail to show the examiner you understand the algorithm.

Starting at A. The nearest town to A is B, 38 km away. C is the town nearest to B, and so on. A → B (38) → C (40) → F (44) → E (32) → D (25) → G (50) → A (48)

Write down the total length of the route.

The total length is 277 km.

EXAM QUESTION 2

Find an upper bound and a lower bound for the length of the optimal salesperson's tour on this network.

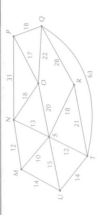

DON'T FORGET
There is more than one way of calculating upper and lower bounds. Be clear how you have found them and why they are an upper or lower bound.

Upper bound

Find a minimum spanning tree. Begin at M. Length of minimum spanning tree is 119. Traversing each edge twice gives 119 × 2 = 238

Lower bound (choose to omit Q)

Remove vertex Q then find a minimum spanning tree for the remaining vertices. To get the lower bound, add the lengths on the minimum spanning tree and the two shortest vertices from Q

$$(12 + 10 + 14 + 12 + 18 + 18 + 17) + (22 + 18) = 141$$

EXAM QUESTION

Here are the tasks from a recipe for making raspberry yogurt ice, along with their duration and the preceding tasks.

		Preceding tasks	Duration (mins)
A	Get out all the ingredients	–	10
B	Sieve raspberries, removing pips	A	5
C	Add sugar, syrup, yogurt to raspberries	B	4
D	Whisk the cream in a bowl	A	6
E	Add whisked cream to raspberry mixture	C, D	3
F	Whisk egg whites and sugar until stiff and white	A	10
G	Mix egg white into raspberry mixture	E, F	5
H	Pour into a freezer container and cover	G	7
I	Freeze	H	240

(a) **Draw a network of this information; include all relevant times.**

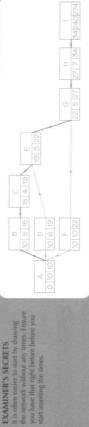

EXAMINER'S SECRETS

It is often easier to start by drawing the network without any times. Ensure you have that right before you start inserting the times.

(b) **Write down the critical path and the time it takes.**

A→B→C→E→G→H→I

It takes 274 minutes

(c) **Which task has the greatest float?**

Task D

(d) **What is the minimum number of people required to make this recipe in the quickest time?**

The minimum number of people is 3

Critical path analysis

Critical path analysis (CPA) helps plan and schedule projects which have many tasks; it minimizes the amount of waiting for earlier tasks to finish before later tasks can begin.

SYLLABUS CHECK

Different labelling methods are used by different boards and teachers. Check which method you are meant to use.

> Here is a table of tasks for making a cup of tea. Complete the second column with the relevant letters. If you think task B has to precede task C, then write B next to C in the 'Preceding tasks' column.

NETWORKS ○○○

A project can be split into tasks. Some of the earlier tasks must be completed before a later task can be started.

	Task	Preceding tasks
A	Water in kettle	none
B	Kettle on	A
C	Cups out	none
D	Milk in cups	C
E	Tea leaves into teapot	none
F	Boiling water into teapot	B, E
G	Pour tea into cups	F, D

DON'T FORGET

When completing the times on a network, keep reminding yourself that it is the *earliest* possible start time and the *latest* possible finish time. Fill in the numbers working forwards and then work backwards adjusting the finish times.

> Add (1) (2) (3) (4) to the correct part of the box, where
> (1) = task or activity
> (2) = earliest possible start time
> (3) = duration of task
> (4) = latest possible finish time

EARLIEST POSSIBLE START AND LATEST POSSIBLE FINISH ○○○

Once the list of tasks is organized, times need to be allocated and a network set up. On a network each task is written into a box like this.

(2)	(1)	(4)
(3)		

Here is a network for erecting and setting up a trailer tent.

From this network we can establish the critical path and the shortest time for completing the project.

A→B→C→F→G→L

or A→B→C→F→K→L

Shortest time = 36 minutes

> Complete the network by putting in all the required times, in minutes.

WATCH OUT

The earliest possible start time plus the duration of the activity will not always give you the latest possible finish time. You then have some float time.

> Write down a possible critical path and its duration.

RESOURCE LEVELLING AND GANTT CHARTS ○○○

A Gantt chart uses thick horizontal lines to represent the project tasks; it plots the days along the top of the chart. The number of people assigned to a task is often written above its line on the Gantt chart. Resource levelling is a way to make projects more cost-effective by graphing number of people against days. The graph is a vertical bar graph and resource levelling tries to reduce the variation between the heights of the bars.

Turn the page for some exam questions on this topic ▶

Linear programming

Linear programming is used in many areas of business and manufacturing. Questions are often about maximizing profit.

LINEAR PROGRAMMING: THREE MAIN COMPONENTS

Variables are values which can change, and ultimately they are what will be solved when finding the optimum values for the problem.

An objective is what the question is asking you to do, i.e. what you are trying to find out; the usual aim is to maximize profits.

Constraints restrict the values the variables can have.

> Give a brief description for each of these terms.

EXAMINER'S SECRETS
Linear programming questions can be quite long. Remember these three main parts and it should become easier to sort out the information.

FORMING THE EQUALITIES AND INEQUALITIES

A factory produces two different types of furniture: chairs and tables. The chairs need 3 hours on the lathe and 1 hour on the sprayer. The tables need 1 hour on the lathe and half an hour on the sprayer. A chair makes a £20 profit whereas a table makes an £8 profit. The lathe is available for use 14 hours per day and the sprayer for 6 hours per day. Total overheads for the factory are £50 per day. How many chairs and tables need to be made per day to maximize profits?

c = number of chairs produced
t = number of tables produced

> Write down the variables.

To maximize profit using the formula
$P = 20c + 8t - 50$ where P is the profit

> Write down the objective.

lathe constraint	$3c + t \leq 14$
sprayer constraint	$c + 0.5t \leq 6$
no negative chairs or tables	$c \geq 0, t \geq 0$

> Write down the constraints.

PLOTTING THE GRAPH

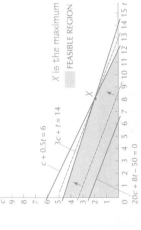

X is the maximum

FEASIBLE REGION

> Using a ruler and keeping it parallel to the line $20c + 8t - 50 = 0$, find the point on the graph that shows the maximum profit.

DON'T FORGET
The maximum profit will be when the ruler is furthest away from the origin and still within the feasible region.

Maximum profit is $P = (20 \times 2) + (8 \times 8) - 50 = £54$ by making 2 chairs and 8 tables.

> Work out the maximum profit and write down how many chairs and tables need to be made.

Turn the page for some exam questions on this topic ▶

For more on this topic, see pages 96–97 of the *Revision Express A-level Study Guide*

EXAM QUESTION

A manufacturer produces Welsh dressers in two types, A and B. Type A requires 4 hours on the lathe, 2 hours to be assembled and 1 hour on the varnisher. It makes a profit of £120 per dresser. Type B requires 2 hours on the lathe, 1 hour to be assembled and $1\frac{1}{2}$ hours on the varnisher. Type B makes a profit of £70 per dresser. The lathe is in operation for 16 hours per day, the assemblers are available for 12 hours per day and the varnisher is available for 6 hours per day. Total overheads are £80 per day. How many of each type should be made to maximize profits?

Variables
A = number of type A dressers
B = number of type B dressers

> Write down the variables, the objective and the constraints.

Objective
To maximize $P = 120A + 70B - 80$

Constraints
$4A + 2B \leq 16$
$2A + B \leq 12$
$A + 1.5B \leq 6$
$A \geq 0, B \geq 0$

> Now draw the graph.

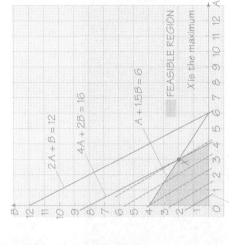

$2A + B = 12$

$4A + 2B = 16$

$A + 1.5B = 6$

FEASIBLE REGION

X is the maximum

$120A + 70B - 80 = 0$

From the graph, maximum profit per day occurs for 3 type A dressers and 2 type B dressers

Maximum profit per day is

$P = 120 \times 3 + 70 \times 2 - 80 = £420$

> Write down the number of type A dressers and the number of type B dressers that need to be made for maximum profit; work out the profit per day.

Descriptive statistics

The tasks in this section are just to consolidate your GCSE data handling. There's still lots to remember, but your calculator will help if you use it carefully.

MEAN, MODE, MEDIAN AND RANGE

Score (x)	1	2	3	4	5	6
Frequency (f)	10	18	8	8	7	9

Modal score = 2 (the most common score)

Range = 6 − 1 = 5 (max score − min score)

Mean = $\sum fx / \sum f$

$$= \frac{(10 \times 1) + (18 \times 2) + (8 \times 3) + (8 \times 4) + (7 \times 5) + (9 \times 6)}{10 + 18 + 8 + 8 + 7 + 9}$$

= 191/60 = 3.2 (1 d.p.)

Median = 3 (30th and 31st values lie in score 3)

Look at this frequency table then calculate the mean, mode, median and range of the scores.

DON'T FORGET $\sum$ means the sum of.

DON'T FORGET When there's an even number of values, the median is the mean of the middle two.

GROUPED DATA

Weight (kg)	2.5 ≤ w < 3	3 ≤ w < 3.5	3.5 ≤ w < 4	4 ≤ w < 4.5	4.5 ≤ w ≤ 5
Frequency	14	40	23	21	17

$\sum fx$ = (14 × 2.75) + (40 × 3.25) + (23 × 3.75)
+ (21 × 4.25) + (17 × 4.75) = 424.8

$\sum f$ = 14 + 40 + 23 + 21 + 17 = 115

Mean = $\sum fx / \sum f$ = 424.8/115 = 3.7 kg (1 d.p.)

From this grouped frequency data, calculate an estimate of the mean.

DON'T FORGET Use class midpoints to help calculate an estimate of the mean.

STANDARD DEVIATION

For individual data

$$\sigma = \sqrt{\frac{\sum (x - \mu)^2}{n}} = \sqrt{\frac{\sum x^2}{n} - \mu^2}$$

For a frequency distribution

$$\sigma = \sqrt{\frac{\sum f(x - \mu)^2}{\sum f}} = \sqrt{\frac{\sum fx^2}{\sum f} - \mu^2}$$

Mean $\mu = \frac{1}{n}\sum x = 90/6 = 15$

Variance $\sigma^2 = \frac{1}{n}\sum x^2 - \mu^2 = (10^2 + 11^2 + 15^2 + 16^2 + 18^2 + 20^2)$
$= 1426/6 - 225 = 12.667$

Standard deviation $\sigma = \sqrt{12.667} = 3.56$ (2 d.p.)

Write out the formulae for calculating standard deviation.

Find the mean and standard deviation of 10, 11, 15, 16, 18, 20.

DON'T FORGET The standard deviation is the square root of the variance.

Turn the page for some exam questions on this topic ▶

For more on this topic, see pages 102, 104 and 108 of the *Revision Express A-level Study Guide*

EXAM QUESTION 1

The ages of the ten people working at an office are 18, 20, 23, 23, 29, 32, 33, 38, 40, 48. Find the mean and the standard deviation of their ages. What was the mean and standard deviation of their ages three years ago?

Mean and standard deviation now

Sum of ages	= 304	
Mean	= 304/10	= 30.4
Sum of squares	= 18² + 20² + 23² + …	= 10 084
Variance	= (10 084)/10 − 30.4²	= 84.24
Standard dev	= √84.24 = 9.18 (2 d.p.)	

Mean and standard deviation three years ago

Three years ago, mean = 30.4 − 3 = 27.4

The spread of ages is no different
so the standard deviation is still 9.18

Remember that standard deviation is a measure of spread. Was the spread of ages different three years ago?

EXAM QUESTION 2

Two squash players counted the number of shots they played in each rally. They recorded their results in a grouped frequency table.

Shots	Rallies
1-10	5
11-20	19
21-30	16
31-40	10
Total	50

Find estimates for (a) the median, (b) the mean and (c) the standard deviation of the number of shots.

Use the extra columns if you wish.

(a) Estimate the median
Draw a cumulative frequency polygon and read off the 25th value, so median = 21

(b) Estimate the mean
Use the extra column or set it out like this

$\sum fx$ = (5 × 5.5) + (19 × 15.5) + (16 × 25.5) + (10 × 35.5)
= 1085

$\sum f = 50$ ∴ mean = 1085/50 = 21.7

(c) Estimate the standard deviation

$\sum fx$ = (5 × 5.5²) + (19 × 15.5²) + (16 × 25.5²)
+ (10 × 35.5²) = 27722.5

Variance = 27722.5/50 − 21.7² = 83.56

Standard deviation = √83.56 = 9.14 (2 d.p.)

DON'T FORGET To estimate the median, draw a cumulative frequency polygon.

S1 Statistical diagrams

OCR-S1 EDEXCEL-S1 AQA(A)-Me WJEC-S1 AQA(B)-S1 CCEA-A2 MEI-S1

This section looks at ways to present raw and grouped data. Just be careful with choice of grouping and class boundaries.

RAW DATA

> Draw a stem and leaf diagram for this data, and from your diagram find the median.

Stem and leaf diagrams display the raw data, but give an idea of spread

17, 9, 11, 23, 12, 20, 13,
22, 14, 11, 21, 8, 19

2 | 3 means 23

```
0 | 8 9
1 | 1 1 2 3 4 7 9
2 | 0 1 2 3
```

Median = 14

CUMULATIVE FREQUENCY

> Complete this table and draw a cumulative frequency polygon — you will need to use graph paper to make sure it's accurate.

> **DON'T FORGET**
> Plot the cumulative frequencies against the upper class boundaries (UCBs)

> Use your cumulative frequency polygon to estimate the median and interquartile range, and draw a box plot using 2.5 as the min and 26.5 as the max.

Cumulative frequency looks at the spread of data. This table gives the heights of seedlings correct to the nearest centimetre.

Height (cm)	Frequency	UCB	Cum. freq.
3–6	10	6.5	10
7–10	19	10.5	29
11–14	49	14.5	78
15–18	59	18.5	137
19–22	27	22.5	164
23–26	16	26.5	180

IQR = 6.7

2.5   11.75 15.25 18.45   26.5

HISTOGRAMS

> If the class widths are different then use frequency density. Complete the table and draw the histogram.

Length (mm)	Frequency	Class width	Freq. density
1–20	5	20	0.25
21–23	12	3	4
24–26	28	3	9.333
27–30	5	4	1.25

Turn the page for some exam questions on this topic ▶

For more on this topic, see pages 106 and 110 of the *Revision Express A-level Study Guide*

EXAM QUESTION 1

> **DON'T FORGET**
> A cumulative frequency polygon is made by joining the points with straight lines.

This table summarizes the weights of 250 people; the weights are in kilograms to the nearest 100 g. Complete the empty columns. Represent the data on a cumulative frequency polygon. Make an estimate of the weight exceeded by 20% of the people.

Weight (kg)	No. of people	UCB	Cum. freq.
44.0–47.9	3	47.95	3
48.0–51.9	17	51.95	20
52.0–55.9	50	55.95	70
56.0–57.9	45	57.95	115
58.0–59.9	46	59.95	161
60.0–63.9	57	63.95	218
64.0–67.9	23	67.95	241
68.0–71.9	9	71.95	250

To find the weight exceeded by 20% of the people means looking at the top 20%, the heaviest 50. Reading from 200 gives about 62.8 kg.

EXAM QUESTION 2

> **DON'T FORGET**
> The boundaries for the first class are 1.5 min and 3.5 min.

> In the last row you must pick a sensible value for the maximum.

A doctor was looking at consultation times. The doctor timed 250 appointments, measured to the nearest minute. Use the table to help construct a histogram for the data.

Time (min)	No. of appointments	Class width	Freq. den.
2–3	30	2	15
4	46	1	46
5	48	1	48
6–7	84	2	42
8–10	27	3	9
11–	15	5–15	3

Collection of data

Why do we collect data and what types can we collect?

POPULATION AND SAMPLE

○○○

Complete the definitions.

Population	a collection of individuals or items
Sample	a selection from the population
Finite population	a population where every member can be given an individual number
Infinite population	a population where it's impossible to number each member
Countably infinite population	an infinite population where each member can be given an individual number

What are the advantages of sampling?

Population too big, cheaper, quicker

What are the disadvantages of sampling?

May not give a true picture of the whole population due to natural variation (chance differences) and bias (non-response, incomplete sampling frame)

Try to list at least two advantages and two disadvantages.

TYPES OF DATA

○○○

Complete the definitions.

Qualitative data is	non-numeric
Quantitative data is	numeric
Discrete data takes	exact values
Continuous data can	have a range of values
Primary data is	data you've collected yourself
Secondary data is	data collected by someone else
Height	quantitative and continuous
Eye colour	qualitative and discrete
Size of family	quantitative and discrete
Cost in pence	quantitative and discrete
Volume	quantitative and continuous

Describe each type of data using one word from each of these two pairs: qualitative and quantitative, discrete and continuous. Just pick one word for eye colour

SAMPLING METHODS

○○○

Link the method on the left with its definition on the right.

1	Random		This involves taking items at regular intervals, e.g. every fifth tree when sampling in a forest.	②
2	Systematic		This is used to ensure the sample is representative of the population; the quota sampling method is often used instead.	③
3	Stratified		This is where each individual must have an equal chance of being chosen.	①

Turn the page for some exam questions on this topic ➤

EXAM QUESTION 1

○○○

Here is the split of year groups in a secondary school. To conduct a survey, they want a sample of 100 students. How many students from each year group are needed in order to get the correct proportional representation?

Year 7	Year 8	Year 9	Year 10	Year 11	Year 12
146	172	158	124	110	70

Think carefully how to get the correct number from each year group.

Total students = 780
For Y7: 146/780 × 100 = 18.7 so 19 students
For Y8: 172/780 × 100 = 22.1 so 22 students
Similarly, Y9 = 20 students, Y10 = 16 students,
Y11 = 14 students, Y12 = 9 students

EXAM QUESTION 2

○○○

A survey is carried out to investigate the quality of new houses. All the people in the country who have recently bought a new house are to be surveyed.

(a) Give one reason why a pilot survey might be carried out first.
(b) Give one reason why the survey should be carried out by post.
(c) Give one possible disadvantage of a postal survey.

(a) To find any problems with the questionnaire
(b) Much cheaper
(c) Small percentage of replies – letters binned

EXAMINER'S SECRETS
Be brief; don't write an essay.

EXAM QUESTION 3

○○○

(a) State a variable about a car that is discrete.
(b) State a variable about a car that is qualitative.

(a) Number of gears, doors, wheels, passengers
(b) Colour, make, model

EXAM QUESTION 4

○○○

(a) A survey carried out inside a newsagent found that 85% of the population buy a newspaper. Why was this a poor sampling method?
(b) A phone poll carried out at 11 am on a Sunday morning revealed that less than 3% of the population regularly go to church. Why was this a poor sampling method?
(c) Some 52% of the population were estimated to watch the six o'clock news each evening after a survey was carried out at a fitness club. Why was this a poor sampling method?

(a) Answers will be biased towards those who do buy newspapers.
(b) Regular churchgoers will be in church when the poll is carried out.
(c) Results will probably be lower than expected, as many will use the fitness club after work and so will not watch the six o'clock news.

(AS) OCR-S1 EDEXCEL-S1 AQA(A)-Me WJEC-S1 CCEA-S1 MEI-S2

Random variables

Understand and interpret notation for random variables.

DISCRETE RANDOM VARIABLES (DRVs)

Let X be the number of heads when two coins are tossed. Find the probability distribution for X and state why this is a DRV.

$p(0 \text{ heads}) = p(TT) = 0.5 \times 0.5 = 0.25$
$p(1 \text{ head}) = p(HT) + p(TH) = 0.5 \times 0.5 + 0.5 \times 0.5 = 0.5$
$p(2 \text{ heads}) = p(HH) = 0.5 \times 0.5 = 0.25$

x	0	1	2
p(X = x)	0.25	0.5	0.25

This is a DRV because it can take only discrete values. Here they are integers but DRVs can also take non-integer values. $(\sum p(X = x) = 1)$

Complete the calculations, then complete the table, and remember to check that the probabilities add up to 1.

DON'T FORGET
The probability distribution is simply a table showing all the probabilities.

EXPECTATION AND VARIANCE OF A DRV

You need to know how to calculate the expected value E(X) and variance Var(X).

$E(X) = \sum xp(X)$ $\{E(X)\}^2 = \sum X^2 p(X)$
$Var(X) = \sum X^2 p(X) - \{E(X)\}^2$

Consider a biased die. Let X be the score shown when the die is thrown. Here is the probability distribution.

x	1	2	3	4	5	6
p(X = x)	0.1	0.1	0.1	0.3	0.2	0.2

Find the expected score and the variance when the die is rolled.

$E(X) = (1 \times 0.1) + (2 \times 0.1) + (3 \times 0.1) + (4 \times 0.3) + (5 \times 0.2) + (6 \times 0.2) = 4$

$Var(X) = [(1^2 \times 0.1) + (2^2 \times 0.1) + (3^2 \times 0.1) + (4^2 \times 0.3) + (5^2 \times 0.2) + (6^2 \times 0.2)] - 4^2 = 18.4 - 16 = 2.4$

Fill in the formulae.

DON'T FORGET
Always define your random variable if it hasn't been defined in the question.

Complete the calculations.

SYLLABUS CHECK
The notation you are used to could well be different. Stick with what you know.

PROBABILITY FUNCTIONS

A probability function is a simple function for calculating probabilities. A spinner can take the values 0, 1, 2, 3, 4 and the probability function for X.

$p(X = x) = \frac{1}{20}(x^2 - 2x + 2)$. Find the probability distribution for X.

When $x = 0$, $p(X = 0) = \frac{1}{20}[(0)^2 - 2(0) + 2] = 0.1$
When $x = 1$, $p(X = 1) = \frac{1}{20}[(1)^2 - 2(1) + 2] = 0.05$

x	0	1	2	3	4
p(X = x)	0.1	0.05	0.1	0.25	0.5

$\sum p(X) = 0.1 + 0.05 + 0.1 + 0.25 + 0.5 = 1$

If this spinner is used at a fairground and you win a prize for spinning less than 3, the probability can be written $F(t) = p(X < t)$.

$p(X < 3) = p(X = 0) + p(X = 1) + p(X = 2) = 0.25$

Complete the table of probabilities and confirm that it's a DRV.

Work out the probability of winning.

Turn the page for some exam questions on this topic ▶

For more on this topic, see page 118 of the *Revision Express A-level Study Guide*.

EXAM QUESTION 1

The discrete random variable X has the following distribution.

x	3	6	9
p(X = x)	$\frac{1}{2}$	2k	k

Find the constant k and the mean value of X.

Probabilities sum to 1, so $\frac{1}{2} + 2k + k = 1$ ∴ $k = \frac{1}{6}$
$p(X = 6) = \frac{1}{3}, p(X = 9) = \frac{1}{6}$
$E(X) = 3(\frac{1}{2}) + 6(\frac{1}{3}) + 9(\frac{1}{6}) = 5$

DON'T FORGET
The probabilities add up to one.

EXAM QUESTION 2

Write a probability distribution for the number of tails when a coin is tossed four times. Find the probability of getting more than two tails.

Let X be the number of tails
$p(X = 0) = p(HHHH) = 0.5 \times 0.5 \times 0.5 \times 0.5 = 0.0625$
$p(X = 1) = p(THHH) + p(HTHH) + p(HHTH) + p(HHHT)$
$= 4 \times (0.5 \times 0.5 \times 0.5 \times 0.5) = 0.25$
$p(X = 2) = p(TTHH) + p(THTH) + p(THHT) + p(HTTH) + p(HTHT) + p(HHTT)$
$= 6 \times 0.0625 = 0.375$
$p(X = 3) = p(TTTH) + p(TTHT) + p(THTT) + p(HTTT)$
$= 4 \times 0.0625 = 0.25$
$p(X = 4) = p(TTTT) = 0.0625$

x	0	1	2	3	4
p(X = x)	0.0625	0.25	0.375	0.25	0.0625

$p(X > 2) = p(X = 3) + p(X = 4) = 0.3125$

First define your DRV then work out the probabilities, and finally draw up a table.

EXAMINER'S SECRETS
Notice the symmetry of the table when $p = \frac{1}{2}$.

LINKS
See the binomial distribution on p. 95.

EXAM QUESTION 3

There are eight coins in a bag, one £1, three 50p, two 20p and two 10p. One coin is drawn at random. Write a probability distribution then calculate the expected amount and the standard deviation.

Let X be the value of the coin taken from the bag

x	10p	20p	50p	100p
p(X = x)	0.25	0.25	0.375	0.125

Using E(X) $= \sum xp(X)$
$E(X) = (10 \times 0.25) + (20 \times 0.25) + (50 \times 0.375) + (100 \times 0.125)$
$= 39p$
Using Var(X) $= \sum X^2 p(X) - \{E(X)\}^2$
$Var(X) = [(10^2 \times 0.25) + (20^2 \times 0.25) + (50^2 \times 0.375) + (100^2 \times 0.125)] - 38.75^2$
$= 811$

$\text{Standard deviation} = \sqrt{811} = 28p$

Discrete probability distributions

Ninety-seven percent of the chocolates produced by a sweet factory are suitable for sale. What is the probability that 10 chocolates will be faulty in a batch of 100?

PERMUTATIONS AND COMBINATIONS

> Write the formulae for ^nP_r and ^nC_r in full.

$$^nP_r = n!/(n-r)! \qquad ^nC_r = n!/(n-r)!r!$$

> How do you know which one to use?

use ^nP_r if order matters otherwise use ^nC_r

> Begin by deciding whether the order is important.

How many ways are there of arranging three letters from MATHS?

Order important so use $^5P_3$ to get 60

How many ways are there to select any 3 people from a group of 9?

Order unimportant, so use $^9C_3$ to get 84

BINOMIAL DISTRIBUTION

> Write the formulae for the binomial distribution then have a go at the question.

For $X \sim B(n, p)$

$$P(X = r) = {^nC_r}p^r(1-p)^{n-r} \qquad E(X) = np \qquad Var(X) = np(1-p)$$

A die is rolled 5 times. What is the probability of getting 2 sixes? Why is the binomial model best here?

The binomial model is best because there are two outcomes (6 or not 6), and the probability of getting a 6 is constant (successive throws are independent). So for $X \sim B(n, p)$

$p = P(6) = \frac{1}{6}$ $q = P(\text{not } 6) = \frac{5}{6}$ $n = 5$ (trials)

$r = 2$ (number of successes required).

Probability $= {^5C_2}(\frac{1}{6})^2(\frac{5}{6})^3 = 0.161$ (3 d.p.)

> If you have time find P(4 or more sixes) = P(4) + P(5);

$P(4 \text{ or more sixes}) = {^5C_4}(\frac{1}{6})^4(\frac{5}{6})^1 + {^5C_2}(\frac{1}{6})^5(\frac{5}{6})^0 = 0.0033$

POISSON DISTRIBUTION

> Write the formulae for the Poisson distribution then have a go at the question.

The Poisson distribution works with events that are randomly scattered and independent of each other. It is written $X \sim Po(\lambda)$ where λ is the mean.

$$P(X = r) = \lambda^r e^{-\lambda}/r! \qquad E(X) = \lambda \qquad Var(X) = \lambda$$

A football team scores an average of 1.2 goals per game. Find the probability that in their next game they score (a) 3 goals, (b) less than their average.

(a) $\lambda = 1.2$, $r = 3$ so $P(X = 3) = (1.2^3 \times e^{-1.2})/3! = 0.087$

(b) less than average means 0 or 1 goals

$P(X = 0 \text{ or } 1) = P(X = 0) + P(X = 1)$

$= 1.2^0 e^{-1.2}/0! + 1.2^1 e^{-1.2}/1! = 0.663$

POISSON APPROXIMATION TO THE BINOMIAL

> Finish the calculations to this question. Ninety-seven percent of the chocolates produced by a sweet factory are suitable for sale. What is the probability that 10 chocolates will be faulty in a batch of 100?

For a large number of trials and small probability, the Poisson distribution provides an approximation to the binomial using $\lambda = np$.

In this question $X \sim B(100, 0.03)$ is approximated by the Poisson distribution using $\lambda = np$. $100 \times 0.03 = 3$, therefore $X \sim Po(3)$

$P(X = 10) = (3^{10} + e^{-3})/10! = 1.6 \times 10^{-2}$

Turn the page for some exam questions on this topic ▶

For more on this topic, see pages 116, 120 and 122 of the *Revision Express A-level Study Guide*

EXAM QUESTION 1

Use the letters from the word EXAMINATION to obtain (a) the number of different arrangements, (b) the number of different arrangements that begin with X and (c) the number of different arrangements that begin and end in A.

> Take care over the repeated letters.

(a) Number of different arrangements

There are 2 As, 2 Is and 2 Ns, so the number of different arrangements $= 11!/2! 2! 2! = 4 989 600$

Note 11! $= {^{11}P_{11}}$ (^nP_r used as order is important)

(b) Number of different arrangements that begin with X

With X at the beginning that leaves 10 letters to arrange with the same repeats, so number of arrangements $= 10!/2! 2! 2! = 453 600$

(c) Number of different arrangements that begin and end in A

With an A at the beginning and the end there are only 9 left to choose from, but this time there are only two pairs. So it's $9!/2! 2! = 90720$

EXAM QUESTION 2

Four fair coins are tossed and the total number of heads showing is counted. Find the probability of obtaining (a) only 1 head, (b) at least 1 head, (c) the same number of heads as tails.

> Use the binomial model.

(a) Now $X \sim B(4, 0.5)$

$P(X = 1) = {^4C_1} \times 0.5^1 \times 0.5^3 = 0.25$

> P(at least 1 head) $= 1 - P(\text{no heads})$

(b) P(at least 1 head)

$= 1 - P(\text{no heads}) = 1 - P(X = 0)$

$= 1 - ({^4C_0} \times 0.5^0 \times 0.5^4) = 0.9375$

(c) P(2 heads) $= P(X = 2)$

$= {^4C_2} \times 0.5^2 \times 0.5^2 = 0.375$

EXAM QUESTION 3

The number of phone calls received at a switchboard on a weekday afternoon follows a Poisson distribution with a mean of 7 calls per five-minute period. Find the probability that (a) there are no calls in the next five minutes, (b) four calls are received in the next five minutes, (c) more than two calls are received between 3:35 and 3:40.

> Start by deciding on the value of λ.

(a) X is the number of calls received in a five-minute period; $\lambda = 7$ so $X \sim Po(7)$

$P(X = 0) = 7^0 \times e^{-7}/0! = 0.0009$ (4 d.p.)

(b) $P(X = 4) = 7^4 \times e^{-7}/4! = 0.0912$ (4 d.p.)

> For part (c) do something similar to Question 2(b).

(c) $P(X > 2) = 1 - \{P(0) + P(1) + P(2)\}$

$P(X = 0) = 0.0009$ from part (a)

$P(X = 1) = 7^1 \times e^{-7}/1! = 0.0064$

$P(X = 2) = 7^2 \times e^{-7}/2! = 0.0223$

$= 1 - \{0.0009 + 0.0064 + 0.0223\}$

$= 0.9704$

Normal distribution

The normal distribution has many examples that you need to learn.

STANDARDIZING DATA

This begins as a simple method of comparison. Consider two test scores: 64 in maths (mean 54 and standard deviation 5) and 78 in biology (mean 68 and standard deviation 8).

$$\text{Maths: } \frac{64-54}{5}=2 \qquad \text{Biology: } \frac{78-68}{8}=1.25$$

Maths was the better (2σ above the mean)

> Standardize these scores using the formula $z=\dfrac{x-\mu}{\sigma}$

> **DON'T FORGET**
> Keep a table of z-values to hand. And remember that they're cumulative.

NORMAL DISTRIBUTIONS

A normal distribution can be written $X \sim N(\mu,\sigma^2)$. By using the formula $Z = (Z - \mu)/\sigma$ the distribution can be standardized to $Z \sim N(0,1)$ then a curve and tables can be used to solve problems for any normally distributed variable.

If $X \sim N(50,8)$ find $P(X<53)$, $P(X<48)$ and $P(53 < X <55)$.

$$P(X<53) = P\left[Z< \frac{53 - 50}{\sqrt{8}}\right]$$
$$= P[Z< 1.06] = \Phi(1.06) = 0.855$$

$$P(X<48) = P\left[Z< \frac{48 - 50}{\sqrt{8}}\right]$$
$$= P[Z< -0.71]$$
$$= P(Z< -0.71)$$
$$= 1 - \Phi(0.71) = 0.239$$

$$P(53<X<45)$$
$$P\left(\frac{53 - 50}{\sqrt{8}} < Z < \frac{55 - 50}{\sqrt{8}} \right)$$
$$= P(1.06 < Z < 1.77) = \Phi(1.77) - \Phi(1.06)$$
$$= 0.961 - 0.855 = 0.106$$

> Fill in the missing items. Standardize then do a sketch graph to help you look up Φ.

> **DON'T FORGET**
> The curve is symmetrical.

BINOMIAL TO NORMAL

If $X \sim B(n,p)$ and n is large so that $np > 5$ and $np(1 - p) > 5$ we can approximate the binomial using the normal.

$$\text{Mean } \mu = np \qquad \text{Variance } \sigma^2 = np(1 - p)$$

If $X \sim B(48, 0.25)$ find the appropriate normal approximation to use.
$$\mu = 12 \text{ and } \sigma^2 = 9 \; \therefore \text{ use } Y \sim N(12,9)$$

> Fill in the formulae that convert the binomial to the normal.
> Have $n = 48$ and $p = 0.25$ so what are μ and σ^2?

CENTRAL LIMIT THEOREM

The central limit theorem (CLT) works with samples (size $n \geq 30$) and the formula is $X \sim N(\mu,\sigma^2/n)$.

If the weights of rabbits follow a normal distribution X such that $X \sim N(450,50^2)$, and if a sample of 40 is taken, what are the mean and variance of the sample?
What is the standard error?
Mean $= 450$
Variance $= 50^2/40 = 62.5$
Standard error $= \sqrt{62.5} = 7.9$

> Standard error is the standard deviation of the sample mean.

Turn the page for some exam questions on this topic ▶

For more on this topic, see pages 126 and 128 of the *Revision Express A-level Study Guide*

EXAM QUESTION 1

Brown sugar is sold in bags with masses which are normally distributed with mean 500 g and standard deviation 4 g. What proportion of bags have a weight between 499 g and 501 g?

Let X = weight of bag then $X \sim N(500, 4^2)$
$$P(499 < x < 501) = P\left(\frac{499 - 500}{4} < Z < \frac{501 - 500}{4} \right)$$
$$= P(-0.25 < z < 0.25)$$
$$= 2\Phi(0.25) - 1$$
$$= 2(0.5987) - 1 = 0.1974$$
$$0.1974 \times 100 = 19.74$$
So 19.7% weigh between 499 g and 501 g

> Standardize the results, sketch a graph then use your tables.

> **DON'T FORGET**
> Finish off the question by interpreting your answer.

EXAM QUESTION 2

Records from a doctor's surgery show that the probability of waiting for more than 15 minutes to go into the surgery is 0.025. If the duration for waiting to go into the surgery is normally distributed with standard deviation of 2.6, what is the mean duration?

Let X be the duration for waiting to go into the surgery, then $X \sim N(\mu, 2.6^2)$
Told $P(X>15) = 0.025$
$\therefore P(X<15) = 1 - 0.025 = 0.975$
$$P(Z< (15 - m)/2.6) = 0.975$$
From tables $0.975 = \Phi(1.96)$
$$15 - \mu = 1.96 \times 2.6$$
$$\mu = 15 - 5.096 = 9.904 \text{ min}$$
Mean duration is approximately 9.9 min

> You'll need to work backwards from your tables this time. To start off, write out everything you know.

EXAM QUESTION 3

It is known that 2% of all batteries are faulty. What is the probability that there will be 20 or more faulty batteries in a batch of 1000?

Let X be the number of faulty batteries.
$\therefore X \sim B(1000, 0.02)$ so $n = 1000$ and $p = 0.02$
$\therefore \mu = np = 1000 \times 0.02 = 20$
$\sigma^2 = np(1 - p) = 20 \times 0.98 = 19.6$
$\therefore X \sim B(1000, 0.02)$ goes to $Y \sim N(20, 19.6)$
Now $P(X \geq 20) = P(Y>19.5)$
$$P(Y>19.5) = P\left(Z > \frac{19.5 - 20}{\sqrt{19.6}} \right)$$
$$= P(Z> -0.11)$$
$$= P(Z< 0.11)$$
$$= \Phi(0.11) = 0.5438$$

> This is a binomial model, so you'll need to approximate to a normal model.

> **DON'T FORGET**
> The binomial distribution is discrete and the normal distribution is continuous. When you standardize, you must apply the continuity correction to $X \geq 20$.

> **IF YOU HAVE TIME**
> Think what the x-value would be if the question read 'more than 20'?

Probability

(AS) OCR-S1 EDEXCEL-S1 AQA(A)-Me WJEC-S1 AQA(B)-S1 CCEA-A2/S1 MEI-S1

Most probability questions can be solved using simple formulae but first of all you need to understand what you're being asked.

ELEMENTARY PROBABILITY

Write in words the meanings of $P(A \cap B)$, $P(A \cup B)$ and $P(A')$, then complete the addition formula.

$P(A \cap B)$ probability A and B happen
$P(A \cup B)$ probability A or B happens
$P(A')$ probability A doesn't happen
$P(A \cup B) = P(A) + P(B) - P(A \cap B)$

If $P(A) = 0.6$, $P(B) = 0.3$, $P(A \cup B) = 0.8$, find $P(A')$, $P(A \cap B)$ and $P(A' \cap B)$.
$P(A') = 1 - 0.6 = 0.4$ $P(A \cap B) = 0.6 + 0.3 - 0.8 = 0.1$
$P(A' \cap B) = 0.2$ as $P(A' \cup B) = 0.2 + 0.3 = 0.5$

MUTUALLY EXCLUSIVE AND INDEPENDENT EVENTS

Explain mutually exclusive.
Mutually exclusive events
A, B cannot happen at the same time; $P(A \cap B) = 0$

Complete the formula.
$P(A \cup B) = P(A) + P(B)$

Explain independent events.
Independent events
A, B have no effect on each other, e.g. flipping a head and rolling a six

Complete the formula.
$P(A \cap B) = P(A) \times P(B)$

EXHAUSTIVE EVENTS

Give two examples of an exhaustive event.
This is where $P(A \cup B) = 1$.
even or odd with dice, red or black with cards

CONDITIONAL PROBABILITY

Let A be taking a red counter second time and B be taking a blue, so P(B) = 3/10.
$P(A \mid B) = P(A \cap B)/P(B)$

If there are 4 red, 3 blue and 3 yellow counters in a bag, and two are taken which are not replaced, find the probability that the second counter was red given that the first counter was blue.

Complete the formula.
Either say there are 4 possibilities out of the remaining 9 counters hence $\frac{4}{9}$, or use the formula

$P(A \mid B) = P(A \cap B)/P(B) = \left(\frac{4}{9} \times \frac{3}{10}\right)/\frac{3}{10} = \frac{4}{9}$

TREE DIAGRAMS

Construct a tree diagram for taking the counters in the above example, then work out the probability of getting a red counter and a yellow counter.

By showing all possible outcomes, tree diagrams make it easier to work out probabilities of combined events, especially when there is more than one way of doing something.

$P(R \text{ and } Y) = P(RY) + P(YR)$
$= \left(\frac{4}{10} \times \frac{3}{9}\right) + \left(\frac{3}{10} \times \frac{4}{9}\right) = \frac{4}{15}$

Turn the page for some exam questions on this topic ►

For more on this topic, see pages 112 and 114 of the *Revision Express A-level Study Guide*

EXAM QUESTION 1

In a certain town 60% of households have a freezer, 75% have a TV and 50% have both. Calculate the probability that a household chosen at random has both a TV and a freezer.

$P(F \cup T) = P(F) + P(T) - P(F \cap T) = 0.6 + 0.75 - 0.5 = 0.85$

EXAM QUESTION 2

A student walks, cycles or gets a lift to school with probabilities 0.2, 0.3 and 0.5 respectively. The corresponding probabilities of being late are 0.3, 0.25 and 0.45 respectively. (a) Find the probability that the student is late on any particular day. (b) Given the student is late one day, find the probability that they cycled. (c) Given the student is not late one day, find the probability that they cycled. Answer to 3 d.p.

Construct a tree diagram then find the probabilities.

Tree diagram:
W 0.2 → L 0.3, NL 0.7
C 0.3 → L 0.25, NL 0.75
Li 0.5 → L 0.45, NL 0.55

(a) $P(L) = P(WL) + P(CL) + P(LiL)$
$= (0.2 \times 0.3) + (0.3 \times 0.25) + (0.5 \times 0.45)$
$= \mathbf{0.36}$

(b) $P(W \mid L) = P(W \cap L)/P(L)$
Have $P(W \cap L) = 0.2 \times 0.3 = 0.06$; $P(L) = 0.36$
so $P(W \mid L) = 0.06/0.36 = \mathbf{0.167}$

(c) $P(C \mid NL) = P(C \cap NL)/P(NL)$
Have $P(NL) = 1 - 0.36 = 0.64$
$P(C \cap NL) = 0.3 \times 0.75 = 0.225$
so $P(C \mid NL) = 0.225/0.64 = \mathbf{0.352}$

EXAM QUESTION 3

Two darts players are throwing darts at the bullseye. The probability that the first hits the bullseye is 0.35 and the probability that the second hits the bullseye is 0.2.
(a) Find the probability that they both hit the bullseye.
(b) Find the probability that only one of them hits the bullseye.
(c) Find the probability that at least one of them hits the bullseye

The events are independent
(a) $P(\text{hit} \cap \text{hit}) = 0.35 \times 0.2 = 0.07$
(b) $P(\text{hit} \cap \text{miss}) + P(\text{miss} \cap \text{hit})$
$= (0.35 \times 0.8) + (0.65 \times 0.2)$
$= 0.41$
(c) $P(\text{at least 1 hit}) = 1 - P(\text{no hits})$
$= 1 - (0.65 \times 0.8) = 0.48$

Correlation

(AS) OCR-S1 EDEXCEL-S1 AQA(B-S1/S3 CCEA-S2 MEI-S2)

This section looks at putting a numerical value on a relationship.

PRODUCT MOMENT CORRELATION COEFFICIENT

$r = S_{xy} / \sqrt{S_{xx} S_{yy}}$

where $S_{xy} = \sum x_i y_i - \frac{1}{n}(\sum x_i)(\sum y_i)$

$S_{xx} = \sum x_i^2 - \frac{1}{n}(\sum x_i)^2$ $S_{yy} = \sum y_i^2 - \frac{1}{n}(\sum y_i)^2$

x	7	12	13	15	17	18	20	25
y	20	28	23	27	36	30	31	35

$\sum x_i = 127$ $\sum x_i^2 = 2225$ $\sum x_i y_i = 3827$
$\sum y_i = 230$ $\sum y_i^2 = 6824$

$S_{xy} = 3827 - \frac{1}{8}(127 \times 230) = 175.8$
$S_{xx} = 2225 - (\frac{1}{8} \times 127^2) = 208.88$
$S_{yy} = 6824 - (\frac{1}{8} \times 230^2) = 211.5$

$r = S_{xy} / \sqrt{S_{xx} S_{yy}} = 175.8 / \sqrt{208.88 \times 211.5}$
$= 0.836 = 0.84$ (2 s.f.)

There's a strong positive correlation as r is near 1.

$r^2 = 0.84^2 = 0.70$ so 70% of the variation in the y-values is in line with the variation of x-values.

SPEARMAN'S RANK CORRELATION COEFFICIENT

$r_s = 1 - \frac{6\sum d_i^2}{n(n^2 - 1)}$

Name	Rank 1	Rank 2	d_i^2
Geoff	4	2	4
Harry	2	3	1
Ian	1	1	0
John	5	5	0
Keith	3	4	1

$6\sum d_i^2 = 6 \times 6 = 36$, $n = 5$ and $n(n^2 - 1) = 120$
so $r_s = 1 - 36/120 = 0.7$

This tells us that the judges agree quite well as r_s is close to 1.

Side notes:

- Fill in the formula for the product moment coefficient r. Make sure you define all the terms.
- Find r for this data. Find values for $\sum x_i, \sum y_i, \sum x_i^2, \sum y_i^2, \sum x_i y_i$. Then calculate S_{xx}, S_{yy} and S_{xy}. Use the S-values in your formula to obtain r.
- **DON'T FORGET** The stronger the correlation, the nearer r gets to 1 or −1 (−1 for a strong negative correlation). The weaker the correlation, the nearer r gets to 0.
- Find r^2 and explain what this value means.
- **DON'T FORGET** The value r^2 gives the variation in percentage terms.
- Spearman's rank correlation coefficient r_s is a measure of how closely one set of rankings matches up with another. Write down the formula for r_s.
- Complete the table, then work out the value of r_s and interpret your answer.
- **WATCH OUT** If the first ranking were to have Harry and Ian joint first place, you would split the positions and give both a ranking of 1.5.
- **DON'T FORGET** The product moment correlation coefficient and Spearman's rank correlation coefficient do have limitations.

Turn the page for some exam questions on this topic ▶

For more on this topic, see pages 138 and 142 of the *Revision Express A-level Study Guide*

EXAM QUESTION 1

The weights of ten people were recorded versus shot-putting distance.

Weight (kg)	80	81	89	93	96
Putting distance (m)	14.6	13.2	14.1	14.7	15.2
Weight (kg)	100	102	110	125	76
Putting distance (m)	16.2	15.0	18.0	16.8	12.3

Calculate the product moment correlation coefficient and percentage variation. Does a heavier person throw further? Comment.

$\sum x_i = 952$ $\sum x_i^2 = 92\,652$ $\sum x_i y_i = 14\,483.2$
$\sum y_i = 150.1$ $\sum y_i^2 = 2278.31$

$S_{xy} = 14\,483.2 - \frac{1}{10}(952 \times 150.1) = 193.68$
$S_{xx} = 92\,652 - (\frac{1}{10} \times 952^2) = 2021.6$
$S_{yy} = 2278.31 - (\frac{1}{10} \times 150.1^2) = 25.309$

$r = S_{xy} / \sqrt{S_{xx} S_{yy}} = 193.68 / \sqrt{2021.6 \times 25.309}$
$= 0.856 = 0.86$ (2 s.f.)

EXAM QUESTION 2

Eight people ran in two races, the 100 m and 1500 m. Here are their times. Calculate to 2 d.p. (a) Spearman's rank correlation coefficient and (b) the product moment correlation coefficient.

	A	B	C	D	E	F	G	H
100 m	11.0	14.3	12.3	12.8	16.0	15.5	11.1	17.2
1500 m	5:03	4:33	4:28	4:55	5:42	5:41	5:10	5:10

	100 m	1500 m
100 m	11.0 14.3 12.3 12.8 16.0 15.5 11.1 17.2	
1500 m	303 273 268 295 342 341 310 310	

	Rank 100 m	Rank 1500 m	d_i^2
A	1	4	9
B	5	2	9
C	3	1	4
D	4	3	1
E	7	8	1
F	6	7	1
G	2	5.5	12.25
H	8	5.5	6.25

$r_s = 1 - (6 \times 43.5)/(8 \times 63) = 0.48$ (2 d.p.)

$\sum x_i = 110.2$ $\sum x_i^2 = 1555.92$ $\sum x_i y_i = 33\,839.8$
$\sum y_i = 2442$ $\sum y_i^2 = 750\,632$

$S_{xy} = 201.25$, $S_{xx} = 37.915$, $S_{yy} = 5211.5$
$r = 201.25 / \sqrt{37.915 \times 5211.5} = 0.45$ (2 d.p.)

Side notes:

- **WATCH OUT** To find the product moment correlation coefficient you need to change the times for the 1500 m race into seconds only.
- First make a ranking table then do the calculations.

Regression

(AS) OCR-S1 AQA(B)-S1 CCEA-S2 MEI-S2

Drawing the most accurate line of best fit on a scatter diagram is very important in order to make accurate predictions.

SCATTER DIAGRAMS

At AS-level, it is important to identify the variables, so you know which is the independent (explanatory) variable and which is the dependent (response) variable. Getting the scale right will avoid needless loss of marks.

Engine size (l)	1.0	1.1	1.3	1.6	1.8	2.0	2.3	2.8	3.0	4.0
Top speed (mile h⁻¹)	90	92	98	95	101	120	115	130	128	140

The independent variable is the engine size.

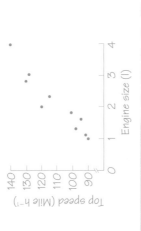

Top speed (Mile h⁻¹) vs Engine size (l)

> **For this data on cars, draw a scatter diagram, stating which of the variables is the independent one**

> **DON'T FORGET**
> The independent variable is x; in this case it's the engine size.

LEAST SQUARES REGRESSION LINE

This is a straight line with equation $y = a + bx$. Look at the engine size data. To calculate the line of regression, you need to find the values of a and b.

> **Complete the steps in the working to find the equation of the regression line. Use the statistical functions on your calculator.**

Step 1 First calculate S_{xy} and S_{xx}.

$S_{xx} = 8.149$

$S_{xy} = 147.09$

> **EXAMINER'S SECRETS**
> It's always a good idea to plot the line on the scatter diagram. If it doesn't fit the data, you know you've made a mistake.

Step 2 Find b using the formula $b = S_{xy}/S_{xx}$.

$b = 147.09/8.149 = 18.05$

Step 3 Find a by substituting the mean values for x and y along with the value of a into the formula $y = a + bx$.

Mean value for $x = \frac{1}{n}\sum x_i = 2.09$

Mean value for $y = \frac{1}{n}\sum u_i = 110.9$

$\therefore 110.9 = a + (18.05 \times 2.09)$ so $a = 73.18$

Equation is $y = 73.18 + 18.05x$

> **WATCH OUT**
> Regression lines have limits. You could even estimate the top speed of a car with a 30 litre engine, which is inconceivable, or you could say a car with no engine size has top speed of 73.18mph.

The regression line can now be used to make predictions – by using the formula or plotting the line on the graph and reading from it.

> **Use your line to predict the top speed of a car with a 10 litre engine.**

Estimate of top speed $= 73.18 + (18.05 \times 10)$
$= 253.68$ mph

Turn the page for some exam questions on this topic ▶

EXAM QUESTION

Engine size (l)	1.0	1.1	1.3	1.6	1.8	2.0	2.3	2.8	3.0	4.0
Average Consumption mile gall⁻¹	48	53	42	40	41	35	27	29	22	19

This table shows the size of ten car engines in litres and their respective fuel consumption in miles per gallon (mile gall⁻¹).
(a) Draw a scatter graph showing this data.
(b) Calculate the least squares regression line and interpret the meaning of a and b. Add this line to your scatter graph.
(c) Use the line to estimate the average consumption of cars with 2.5 litre and 6.8 litre engines. State any problems with these estimates.

(a) Draw the graph.

Average consumption (mile gall⁻¹) vs Engine size (l)

The independent variable x is the engine size.

(b) Calculate the regression line.

Regression line has the equation $y = a + bx$.

Step 1
$S_{xx} = 8.1$ and $S_{xy} = -90.00$

Step 2
$b = -90.0/8.15 = -11.04$

Step 3
Mean value for $x = \frac{1}{n}\sum x_i = 2.09$

Mean value for $y = \frac{1}{n}\sum y_i = 35.6$

Substituting into the equation $y = a + bx$ gives
$35.6 = a + (-11.05 \times 2.09)$ so $a = 58.67$
Equation of regression line is $y = 58.67 - 11.04x$

As b is negative, there is a negative correlation. When $x = 0$, $y = b = 58.67$; this is misleading as you wouldn't have a car without an engine and it wouldn't do 58.67 mile gall⁻¹.

(c) Estimate the average consumption for the two engine sizes.

If $x = 2.5$, $y = 58.67 - (11.04 \times 2.5) = 31.07$ mile gall⁻¹

If $x = 6.8$, $y = 58.67 - (11.04 \times 6.8) = -16.47$ mile gall⁻¹

The figure -16.45 mile gall⁻¹ is obviously in the absurd region. The consumption would be low but not that low.

> **WATCH OUT**
> Use sensible scales for the axes on your scatter diagram.

> **DON'T FORGET**
> Before you start entering all the data on your calculator, make sure you clear the memory of the last items you entered.

> **Comment on your answers once you've worked them out.**

> **Comment on your answers once you've worked them out.**

Index

Index